Nanoscale Microwave Engineering

FOCUS SERIES

Series Editor Pascal Maigné

Nanoscale Microwave Engineering

Optical Control of Nanodevices

Charlotte Tripon-Canseliet
Jean Chazelas

iSTE

WILEY

First published 2014 in Great Britain and the United States by ISTE Ltd and John Wiley & Sons, Inc.

ISTE Ltd
27-37 St George's Road
London SW19 4EU
UK

www.iste.co.uk

John Wiley & Sons, Inc.
111 River Street
Hoboken, NJ 07030
USA

www.wiley.com

Library of Congress Control Number: 2014930265

British Library Cataloguing-in-Publication Data
A CIP record for this book is available from the British Library
ISSN 2051-2481 (Print)
ISSN 2051-249X (Online)
ISBN 978-1-84821-587-0

Contents

Introduction

I.1. General introduction

When addressing the main requirements of future interconnected system environments, autonomy and resilience are the most challenging factors since they include very demanding technology aspects and integrated intelligence aspects.

In an environment where communications often take a larger place, it is obvious that all the future systems will be required to have the capability of working in a networked ambient environment.

New systems will be defined by key words representing their main functions: smart, autonomous, wireless, networked and sensing systems. The adjective smart and autonomous refer to the autonomy in terms of energy, such as zero power consumption or energy harvesting, and to the autonomy defined from the decision point of view (i.e., they have the capability to do what is needed to be done, when it seems optimal to do it).

It also requires the inclusion of all the elements required for a dedicated mission: protection of the environment, communication, security or defense, biomedical and e-health, and power electronics.

I.2. Definition of a new area "nanoarchitectronics"

Richard Feynman's visionary speech in 1959 had inspired the field of nanotechnology, with the theme "to synthesize nanoscale building blocks with precisely controlled size and composition, and assemble them into larger structures with unique properties and functions". Never before in history has any technology provided so many possibilities to create and manipulate such tiny structures as the basic elements for functional devices and hierarchical systems that render superior performances.

Microwave systems, technology and material-based architectures at nanoscale lead to a novel approach and a novel scientific area. We propose to call this new area *nanoarchitectronics* because it describes the ability to build up or design new architectures at the material level, device level and system level, including electromagnetics and electronics at nanoscale, nanomaterials and nanotechnologies based on basic physics and embedded software systems.

As an example and following the research on two-dimensional (2D) atomic crystals, it appears extremely powerful to assemble isolated atomic planes into designer heterostructures made layer by layer in a precisely chosen sequence. The first, already remarkably complex, such heterostructures (often referred to as "Van der Waals") have recently been fabricated and investigated, revealing unusual properties and new phenomena.

In this book, we will address an additional new field, which could be entitled beyond nanoarchitectronics, in which we are looking to a new dimension of the above-mentioned architectures, when dealing with the interactions of electromagnetic waves and nanodevices. As an example of these new capabilities, this book will present new approaches linked to the use of photonics technologies to control nanoscale microwave devices.

We have chosen to reduce the scope of our analysis to the impact of nanotechnologies on electromagnetic (EM) applications ranging from radio frequency (RF) to terahertz (THz) and to extract technologies that exhibit advanced or new performances with size reduction either by continuity like More Moore or by breakthrough quantum effects [PIE 10a].

Interesting nanostructured materials, devices and systems already constitute research areas in RF nanoelectronics [PIE 10b].

I.3. Three main thrusts

I.3.1. *Thrust 1 – around new nanodevices and systems*

The following research fields are concerned: carbon nanotubes (CNT), graphene and nanowire (NW), graphene nanoribbon (GNR) circuits and transmission lines, semiconductor- and other novel material-based nanotechnology for RF electronics, nanostructured microwave materials and metamaterials, nanowireless sensors and power meters, nanoantennas and arrays, THz nanoelectronics/photonics, including signal generation and processing, photoemission and detection, nano-interconnects for advanced RF packaging, nanoscale electromechanical switches (NEMS) and resonators, spin waves for RF nanoelectronics (spintronics) and molecular electronics, nanoplasmonic structures for RF applications and superconducting nanostructures and RF nanodevices for quantum information processing.

I.3.2. *Thrust 2 – around theoretical issues and modeling*

The following research fields are concerned: multiphysics modeling of nanostructures and nanodevices, ballistic transport, wave solutions and multiport circuits in nanomaterials, combined electromagnetic/coherent transport problem in nanodevices, electrodynamics, radiation, detection and photogeneration in nanostructures, and wave mixing, dispersive and nonlinear effects in nanomaterials.

I.3.3. *Thrust 3 – around technology, instrumentation, imaging and reliability*

The following research fields are concerned: broadband characterization of nanoscale devices/systems for RF applications, microwave nanoscale near-field imaging and surface patterning, noise measurement of nanoscale devices and three-dimensional (3D) integration of carbon- and silicon/semiconductor-based nanodevices.

The scientific theories associated with these research areas cover the following fields [ENG 07]:

– advanced metamaterials, electromagnetic/photonic band gap structures;

– nanostructured materials modeling;

– nanoelectromagnetism;

– nanoscale devices and components;

– miniaturized antennas and optical nanoantennas;

– 2D and graphene metamaterials, graphene photonics;

– metatronics: nanocircuits and nanosystems.

This book will focus on the engineering of nanomaterials for microwave, millimeter wave and terahertz applications and especially on the optical control of these nanodevices.

The purpose of this book is to provide the readers with required knowledge to enter the world of nanoarchitects for microwave nanosystems.

Here, we recall the main trends for 3D, 2D and one-dimensional (1D) materials, which could be used in the definition of new system architectures (see [ALA 09, Figure 1]). In the field of the interaction between light and semiconductor materials, some elements of the history of photoconductivity and the capability to go from the picosecond (ps) domain to the THz frequency range will be given.

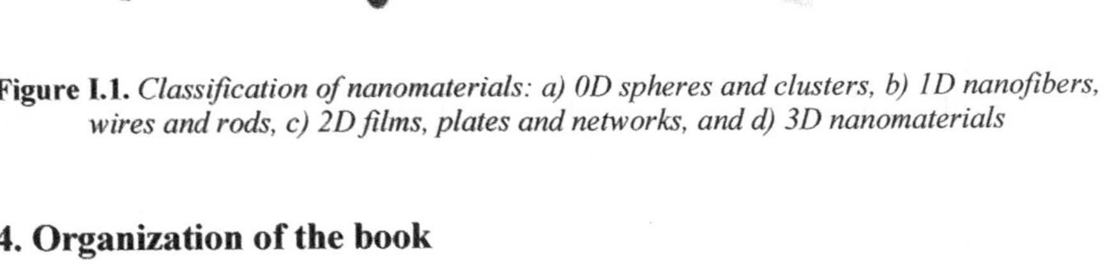

Figure I.1. *Classification of nanomaterials: a) 0D spheres and clusters, b) 1D nanofibers, wires and rods, c) 2D films, plates and networks, and d) 3D nanomaterials*

I.4. Organization of the book

The book is organized as follows: Chapter 1 deals with nanotechnology-based materials for ultrafast microwave applications and their interaction with light. Chapter 1 is focused on two aspects: first to give some trends in new semiconductor materials from 3D to 0D and second to give an in-depth analysis of the interaction at nanoscale between light and these new materials in photoconductivity and plasmonics. The materials concerned are carbon-based materials (especially graphene and carbon nanotubes), NW-based technologies: Si, III–V semiconductors, ZnO, etc., nanostructured materials and metamaterials.

Chapter 2 addresses EM material characterization at nanoscale, including a state of the art of macroscopic material characterization techniques in the microwave domain with dedicated equipment, EM environment constraints (T°, mechanical stability and multiscale access) and noise contributions in measurements and the evolution of techniques for mesoscopic nanomaterial characterization.

Chapter 3 is devoted to nanotechnology-based components and devices, reviewing the existing components and a state of the art with these technologies (active) and with a focus on new passive components and devices with optional optical control (photoconductivity and plasmonics).

Chapter 4 presents the engineering of new optically controlled microwave functions based on 2D and 1D semiconductor materials.

Finally, in the Conclusion, we draw some perspectives of this new field of optically controlled low-dimensional materials with a focus on the so-called Van der Waals heterostructures as an example of nanoarchitectronics.

1

Nanotechnology-based Materials and Their Interaction with Light

Chapter 1 will be dedicated to nanotechnology-based materials for ultrafast microwave applications and the interactions of these materials mainly semiconducting with light. It will focus on two aspects, the first aspect is to give some trends in new semiconductor materials from three dimensional (3D) to zero dimensional (0D) and the second aspect is to give a deep analysis of the interactions at nanoscale between light and these new materials around photoconductivity and plasmonics. Materials concerned are carbon-based materials (especially graphene and carbon nanotubes), nanowire-based technologies: Si, III-V semiconductors, ZnO, nanostructured materials and metamaterials.

1.1. Review of main trends in 3D to 0D materials

1.1.1. *Main trends in 3D materials for radio frequency (RF) electronics and photonics*

Controlling the permittivity and permeability of three-dimensional (3D) materials appears as a major challenge for future electromagnetism systems. Nanomaterials are high-potential candidates for applications in microwave, millimeter wave, terahertz (THz) and optical systems. During the last decade, numerous research activities have been devoted to the study of artificial materials, such as metamaterials [ENG 06]. Mixing components at the nanoscale results in materials providing superior properties compared with conventional microscale composites and, at the same time, that can be synthesized using simple and inexpensive techniques.

In particular, major research advances have been obtained by the group of Nader Engheta, who worked on specific materials such as epsilon-near-zero (ENZ), mu-near-zero (MNZ), zero-index metamaterials and double-negative materials [ALU 07].

The introduction on the structure of the split-ring resonator (SRR) by J. Pendry in 1999 opened the way for the demonstration of metamaterials based on the periodic implantation of such SRRs [PEN 99, PEN 07, SMI 00]. The implementation of the first effective medium with left-handed properties by D.R. Smith in 2000 was possible due to the use of small metallic resonators, SRRs. The SRR appeared as the first non-magnetic resonator capable of showing negative values of the magnetic permeability around its resonance frequency.

The control and the tunability of those materials remain as the great future challenges in this field.

1.1.2. *Main trends in 2D materials for RF electronics and photonics*

The combination of the unique properties of two-dimensional (2D) semiconductor materials, such as graphene, with new device concepts and nanotechnology can overcome some of the main limitations of traditional electronics in terms of maximum operating frequency, linearity and power dissipation.

1.1.2.1. *The example of graphene*

Graphene is a flat monolayer of carbon atoms forming a 2D honeycomb lattice. Graphene is a basic building block of graphite and carbon nanotubes (CNTs). Graphene properties were first introduced by Wallace in 1947.

At the beginning of the 21st Century, Andre Geim, Konstantin Novoselov and their collaborators from the University of Manchester (UK), and the Institute for Microelectronics Technology in Chernogolovka (Russia), published their results on graphene structures in October 2004 [NOV 04].

After reviewing some important papers in the literature devoted to this new material, we can derive some basic characteristics of graphene materials and their main applications.

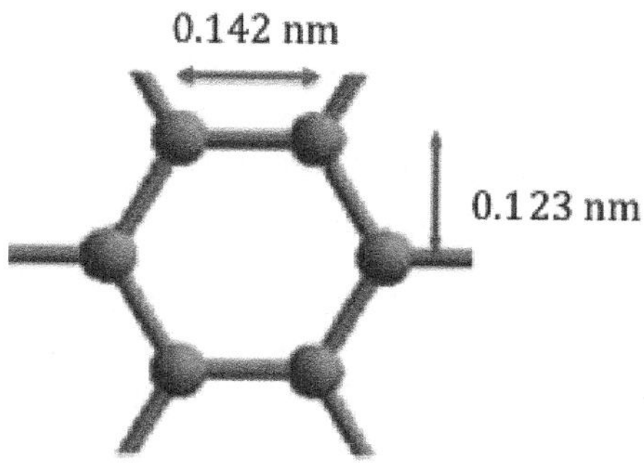

Figure 1.1. *Graphene: a flat monolayer of carbon atoms for ming a 2D honeycomb lattice*

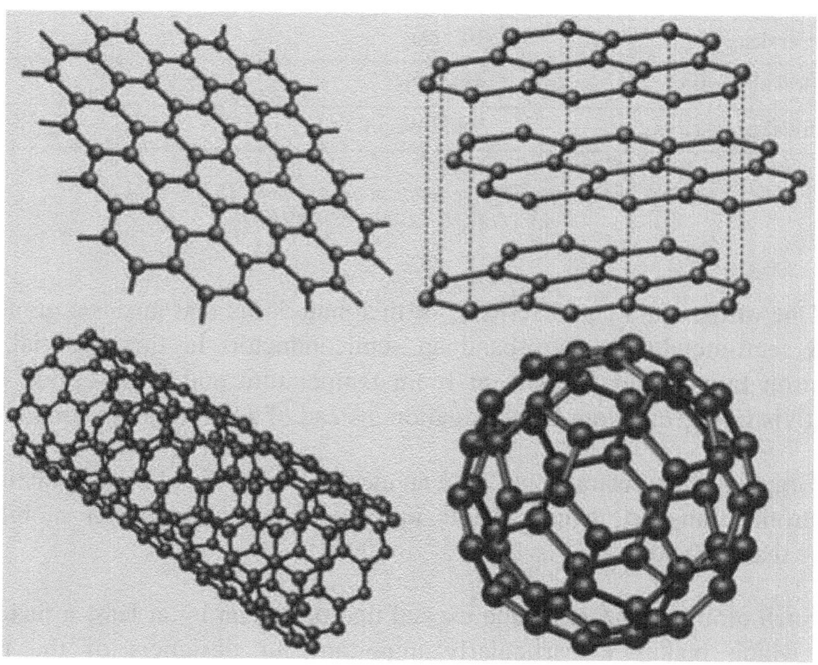

Figure 1.2. *Top left: graphene is a honeycomb lattice of carbon atoms. Top right: graphite can be viewed as a stack of graphene layers. Bottom left: carbon nanotubes are rolled-up cylinders of graphene. Bottom right: fullerenes C60 are molecules consisting of wrapped graphene by the introduction of pentagons on the hexagonal lattice [CAS 06]*

Parameters	Typical characteristics
Thickness	0.142 nm
Band structure	Semi-metal or zero-bandgap semiconductor
Electron transport	Ballistic at room temperature Relativistic quantum Dirac equation
Carrier mobilities	1,00,000 $cm^2/V.s$ in suspended graphene 10,000 $cm^2/V.s$ in graphene on substrate
Young's modulus	1.5 TPa
Breakdown current	10^8 A/cm^2
Current density	1 A/µm
Transistor cutoff frequency	350 GHz
Carrier density	10^{14} cm^{-2}
Optical absorption	$\pi\alpha$~2.3%
Thermal conductivity	5,000 W/m.K

Table 1.1. *Typical characteristics of graphene [CAS 09, AVO 10, WAN 10, LOV 12, WU 12, SCH 10]*

One of the main characteristics of this material is that intrinsic graphene is a semi-metal or a zero-bandgap semiconductor. In this material, the electron transport is ballistic at room temperature and is described by a relativistic-like quantum Dirac equation instead of a Schrödinger equation.

Graphene demonstrates not only an electric field effect but also a ballistic electronic transport, which results in very high charge carrier mobilities more than 100,000 $cm^2/V.s$.

Such mobilities of graphene exceed that of silicon by at least a factor of 40, which makes it particularly important for designers of the next-generation fast transistors.

Also, graphene has a Young's modulus of 1.5 TPa.

Due to these unique properties, graphene is very promising for high-frequency nanoelectronic devices, such as oscillators and switches. In practical applications, graphene is deposited on a SiO_2 layer with a typical thickness of 300 nm, which is grown over a doped silicon substrate.

It is interesting to note that the conductivity of the graphene sheet is an anisotropic tensor and it can be controlled by applying an electrostatic and magnetostatic biasing field. This property introduces the possibility of developing new applications, which cannot be obtained by conventional conducting materials of fixed conductivities.

1.1.2.2. *Graphene for RF applications*

Recent results on the use of graphene for microwave applications enabled us to review some functions covering the field of nanocircuits up to the realization of new microwave functions based on this material; among those, we will review the following functions in Chapter 2: RF mixers, frequency multipliers, antennas, isolator, circuits, transistors and field-effect transistors (FETs), photodetectors, barristor, optoelectronic functions such as graphene photodetector and other potential applications of graphene, superconducting FETs and room-temperature spintronics, and transparent electrodes [OBE 11].

Some helpful analysis on the use of graphene for microwave applications will be given in Chapter 3.

1.1.3. *Review of other two-dimensional structures for RF electronic applications*

1.1.3.1. *Plasmonic structures*

Plasmonics is based on the interaction process between an electromagnetic radiation and the conduction electrons at metallic interfaces or in small metallic nanostructures. For noble metals such as Ag and Au, the plasma frequency is in the visible or ultraviolet region; therefore, their permittivity has negative real parts in the optical frequencies. These metals behave as plasmonic materials, and their interaction with optical signals involves surface plasmon resonances (SPRs). These plasmonic structures provide interesting possibilities not only for synthesizing subwavelength cavities or new metamaterials at infrared and optical frequencies, but also for addressing new microwave functions involving confined optical interactions and microwave modulation of the electric field at a semiconductor/dielectric interface [AHM 12].

A detailed approach of plasmonic structures has been given in the Introduction.

1.1.3.2. *Two-dimensional semiconductor materials*

The 2D semiconductors such as transition metal dichalcogenides (such as MoS_2, $MoSe_2$, WS_2 or WSe_2) show excellent device characteristics, as well as novel optical, electrical and optoelectronic characteristics due to quantum size effects. Recent research of 2D materials based on chalcogenides and/or III–V semiconductors on Si/SiO_2 substrates has been achieved. It is important for both fundamental science and applications, such as electronics, photonics and chemical sensing. Unlike the zero-bandgap graphene, it is possible to tune the bandgap of 2D semiconductor materials by the choice of elements and the number of layers. The large bandgaps of 2D semiconductors (e.g. 1.8 eV for MoS_2 monolayer nanosheet) and their carrier mobility make these materials very attractive for the next-generation nanoelectronic and nanophotonic devices [JAV 13].

1.1.4. *Main trends in 1D materials for RF electronics and photonics*

This section deals with recent research on carbon-based and non-carbon-based one-dimensional (1D) materials such as nanorods/carbon nanotubes (CNTs), boron nitride nanotubes (BNNTs) and semiconductor nanowires (NWs).

1.1.4.1. *CNT materials and microwave applications*

It is well known that CNTs are cylinders of nanometer (nm) diameter of a graphene sheet wrapped up to form a tube. Since their experimental discovery in 1991 [IIJ 91], numerous research efforts have been devoted to exploring their physical properties including electromagnetic wave interaction of the conducting CNTs, which seems to contain important features compared with traditional conductors [GHA 11].

Single-wall carbon nanotubes (SWCNTs) can be seen as rolled-up sheets of graphene, i.e. a monolayer hexagonal lattice of carbon atoms. They can be uniquely described by a double index or chiral vector (n,m) that corresponds to the way the graphene sheet is rolled up, i.e. the angle it makes with the vectors of the lattice and how tightly the CNT is rolled up. According to this index, an SWCNT may be metallic (i.e. no bandgap) or semiconducting (i.e. diameter-dependent bandgap). The only truly metallic CNTs are the armchair n = m, rolled along vector a_2 as shown in Figure 1.3. For zigzag (rolled along vector a_1) and chiral SWCNTs, there are two cases: if $n - m = 3j + 1$ or $3j + 2$, then the CNTs are semiconducting. However, if $n - m = 3j$, the bandgap is sufficiently small in that they behave as metals at ambient temperature. They are semi-metallic tubes, usually referred to as being simply metallic CNTs. In this case, from Figure 1.3, or the three-congruence, it is simple to conclude that, if all chiral vectors have the same probability, there should be 1/3 metallic SWCNTs and 2/3 semiconducting. Common CNT production methods usually have a narrow diameter distribution but chiral vectors of semiconductor and metal SWCNTs are equivalently distributed.

Multiwall carbon nanotubes (MWCNTs) are concentric shells like Russian dolls made up of SWCNTs. They are available in a variety of diameters, number of shells, etc.

CNTs can display huge length-to-diameter aspect ratios because the diameter varies from 1 nm to a few tens of nm, but lengths up to half a meter have been reported [CHA 08]. Usual lengths vary from less than 1 µm to 100 µm depending on the fabrication technique and application.

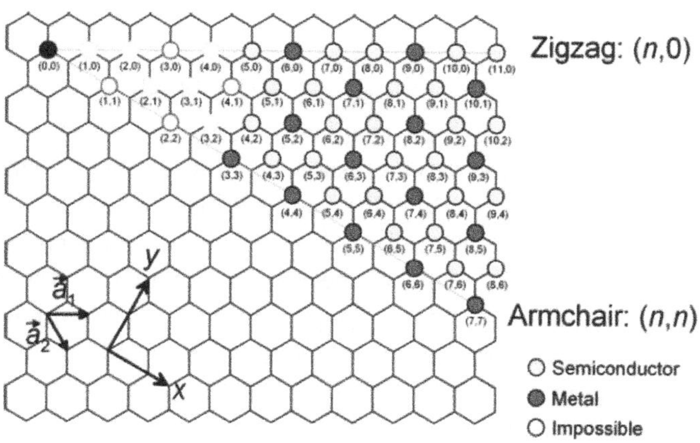

Figure 1.3. *Chiral vector of CNT*

The earliest research has been focused on the synthesis of CNTs necessary for experiments. The progress of common growth techniques including arc discharge and laser ablation of graphite pieces and later chemical vapor deposition (CVD) and plasma-enhanced CVD has significantly improved the quality of the CNTs [MEY 05]. Recently, many companies have concentrated their efforts to develop and control the growth techniques to perform very high purity and uniform CNTs in length and diameter. The physical properties of CNTs [SAI 98] have demonstrated unique electronic structure depending on their dimensions and chiralities, which have been considered as a major discovery in the field of semiconductor devices. In addition, due to strong covalent carbon–carbon bonding, CNTs present very high tensile strength, thermal and electrical conductivity. Hence, those mechanical and electrical properties make them also very attractive for reinforcing composite materials for aerospace applications. Usually, CNTs are classified into two categories: SWCNTs that are composed of a single graphitic cylinder, where the diameter varies from 0.7 to 2 nm, and MWCNTs that are composed of several concentric graphitic layers, where the diameter varies from 10 to 200 nm. Commercial CNTs are available in a powder form where they are entangled and randomly distributed [ZHA 13].

1.1.4.1.1 Carbon nanotubes for RF applications

CNTs are competitive elements in many RF applications like high-frequency transistors, AM demodulators, matched loads, transmission lines and antennas.

The knowledge of dielectric properties at low frequency (direct current (DC) to 1 MHz) as well as high frequency (above 1 MHz) is the key for developing novel electronic devices such as radio frequency-miniaturized resonators [SAB 09a, SAB 09b] as electronic model for future implementation [SAB 09c]. Therefore, the electrical characterization at radio frequency/microwave (RF/M) bands becomes a challenge where common measurement techniques are inappropriate. For high-frequency measurements, complex permittivity is usually used to describe the dielectric properties of a material where the real part of the permittivity is related to the dielectric constant and the imaginary part can be associated with the conductivity [TRI 14].

1.1.4.2. *BNNT materials and microwave applications*

Theoretically predicted in 1994, BNNTs appeared as a 1D allotrope of a 2D sheet of boron nitride in a hexagonal lattice with alternating boron and nitride atoms [RUB 94, COH 10].

Similar to the CNTs, BNNTs have many unique properties. In particular, BNNTs are semiconductors with a wide bandgap (~5.5 eV) weakly dependent on the tube diameter, helicity and the number of walls. Compared with the CNTs, BNNTs have not only high thermal conductivity but also high oxidation resistivity, as well as high thermal and chemical stability, which render BNNTs a promising tubular material for developing nanotube-based electronic devices in certain hazardous and high-temperature environments [WAN 09, WAN 10].

1.1.5. *Other 1D materials for RF applications*

Semiconductor NWs have drawn enormous research efforts for their special material properties and wide-ranging device applications with outstanding performances [SAB 09c, TRI 11, RUB 04]. As one of the elementary building blocks of nanoelectronic devices, NWs configured as FETs have been shown to operate at ultralow power below microwatts with an enhanced operational speed. Semiconductor NWs, such as groups IV [QIN 08, HOC 08], III–V [CHA 08, CUI 03] and II–VI [XIA 06, BRY 06], and others [NIL 08, LIU 05, COM 05] have demonstrated the prospective construction of state-of-the-art devices.

This section deals with recent research on non-carbon-based 1D materials such as semiconductor NWs.

1.1.5.1. *Semiconductor nanowires Si and II–V compounds*

Semiconductor NWs have been extensively investigated as nanoscale building blocks for novel nanoelectronics and nanophotonics. Both top-down fabrications based on conventional microfabrication technologies and top-down fabrication based on chemical synthesis (from the gaseous or from the liquid phase) are possible. In addition to being driven by the need to overcome some limitations of the top-down fabrication, the study of bottom-up NW growth is motivated by their interesting electrical transport and optical properties that are remarkably different from those of the corresponding bulk materials. Homogeneous NWs can be developed by using a number of chemical methods, among which the metalorganic chemical vapor deposition (MOCVD) is arguably the most common for III–V compounds. Apart from homogeneous NWs, it is possible to synthesize heterostructures along the axial and radial directions, which allows for the manipulation of the NW properties and bandgap engineering.

By applying heteroepitaxy on NWs, junction geometries can also be engineered. In addition, it is easier to overcome the lattice mismatch problem when doing heteroepitaxy on bulk and/or using tensile or compressive strain, allowing more flexibility in matching the absorption to the desired wavelength. As already mentioned, NWs are 1D or quasi-1D nanostructures with unusual electronic and optoelectronic properties, arising from unique characteristics such as:

– extremely high aspect ratio: with lengths ranging from ~1 to 10 µm and diameters ranging from ~2 to 100 nm, NWs are ideal for dense device integration and exploitation of size effects, i.e. large surface-to-volume ratio and 2D quantum confinement, which lead to interesting new phenomena;

– synthetic flexibility: both top-down and bottom-up fabrication methods and growth techniques are applicable to the entire range of semiconductors, allowing us to tailor NW optical and the electronic properties through the choice of materials, doping and bandgap engineering;

– NW technologies are in principle compatible with mainstream (CMOS) fabrication processes: top-down methods include *in situ* fabrication by conventional lithographic techniques and processing, or direct transfer of fabricated NWs on a host substrate, similar to the bottom-up methods where additional functionalities may also be integrated during the synthesis through doping or the formation of axial and radial heterostructures.

Moreover, the small lateral dimensions of NWs allow extreme heteroepitaxial growth which may be important, for instance, for the integration of III–V semiconductors on silicon platforms.

1.1.5.2. *Si and GaAs nanowire field-effect transistors*

Single-electron Si NW FETs were fabricated, and a Coulomb blockade was observed even at room temperature ([SUN 11, Figure 4]). The effects of length-induced strain on the transport properties of such devices were investigated.

Gallium arsenide (GaAs) NW FETs with universal (back) gates are routinely fabricated by photolithography and electron-beam lithography on Si/SiO_2 substrates to determine NW transport parameters (Figure 1.5). Transport properties of three-branch monolithic GaAs NW junctions are being investigated for the realization of Y-junction transistors and quantum junction devices [DAI 11].

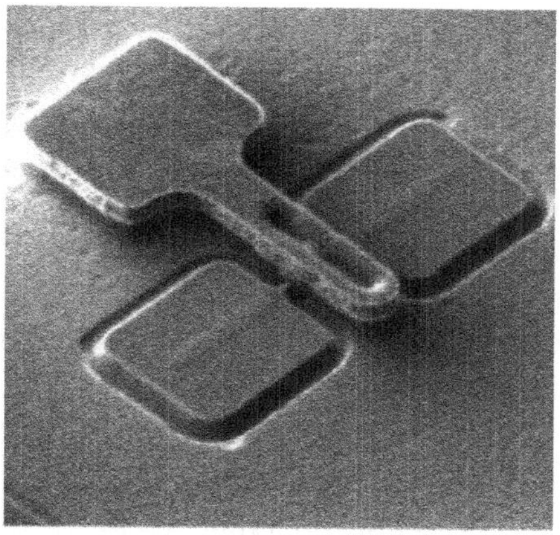

Figure 1.4. *Example of a single-electron Si NW FET realized at the CINTRA UMI*

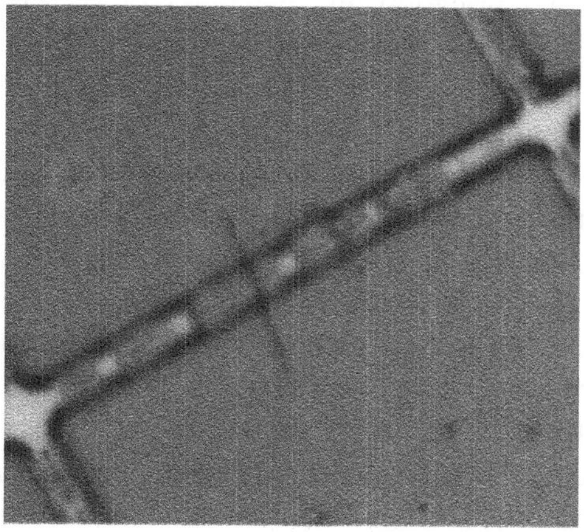

Figure 1.5. *Example of a GaAs NW FET realized at the CINTRA UMI*

NW-based FETs operating as high-sensitivity THz detectors have been realized [VIT 12] with a photovoltage signal corresponding to responsivity values >10 V/W, at 1.5 THz. The potential scalability to even higher frequencies and the technological feasibility of realizing multipixel arrays coupled with quantum cascade laser quantum cascade laser (QCL) sources make the proposed technology highly competitive for a future generation of THz detection systems.

Semiconductor NWs are an ideal building block for implementing rectifying diodes or plasma-wave detectors that could be well operated into the THz, due to their typical attofarad-order capacitance. As active channel of our FET detectors, we select indium arsenide (InAs) NWs since they show reasonably high electron mobility even at room temperature (~1000 cm^2/Vs) and a potentially long electron mean free path, enabling high transconductance at very low drive voltages. InAs NWs with 1.5 m length and having a diameter of 30 nm were grown bottom-up on InAs (111) B substrates by chemical beam epitaxy (CBE).

1.1.5.3. *ZnO nanorods*

In recent years, 1D semiconducting nanostructures have attracted tremendous interest for their unique physical properties attributed to their small dimensions. Driven by their enticing potential as nanoscale building blocks for integrated electronic and photonic circuits, many efforts have been devoted to the fabrication and characterization of functional devices based on 1D nanostructures, such as FETs, sensors, lasers and solar cells. In parallel, integration of device elements has been explored by both top-down and bottom-up techniques. To fully utilize the scaling advantage of the 1D structures, vertical alignment provides an efficient and flexible way to construct 3D architectures. In fact, 3D architecture using semiconducting NWs as scaffolds has been used for vertical FETs [NG 04] and field-emission devices [GAN XX]. In most of the earlier reports, vertical 1D semiconductor materials were grown on epitaxial substrates. The control of their horizontal ordering and density requires a complex process [GAN 04, MAR 03]. In addition, the resulting 1D arrays are free-standing, thus posing difficulty to fabricate "bottom" and "top" electrodes to address individual 1D channels. In this work, a high-density zinc oxide (ZnO) NW array was assembled into an anodic alumina membrane (AAM) via the CVD method assisted by electrodeposition of tin (Sn) catalysts. The electrical transport and photoconduction of individual vertical ZnO NWs were characterized using conductive atomic force microscopy (AFM). AAM has been widely used as a template for fabrication and direct assembly of a variety of 1D nanostructures [GAN 04]. Its advantage lies in the convenience of controlling the aspect ratio and the integration density.

Among the II–VI semiconductors, ZnO NWs have been extensively studied for their abundant physical properties and numerous device applications. At the material level, ZnO has a stable wurtzite crystal structure. The zinc atoms are tetrahedrally coordinated with four oxygen atoms, which can be considered as two interpenetrating hexagonal lattices of zinc and oxygen. ZnO is a semiconductor that shows strong piezoelectric and pyroelectric properties. It has a direct wide bandgap of 3.37 eV at room temperature. As a result, ZnO nanostructures have become cutting edge nanotechnology research in transparent electronics, actuators, sensors, optoelectronics and spintronics, especially ZnO NW-based electronic devices including FETs [CHA 08].

1.1.6. *Some attempts on 0D materials*

Earlier, it was shown that quantum dots (QDs) deposited within or over GaAs can enable and/or enhance the efficiency of THz signal generation [LEY 09]. Here we describe the efficient generation of THz output signals using PC THz antennas based on semiconductor structures comprising InAs QDs embedded in high-quality crystalline GaAs, whereby the embedded QDs act as the ultrafast capture mechanism [EST 09, RAF 04].

1.2. Light/matter interactions

This section addresses the interaction between a light wave and a material aiming at the realization of new optically controlled microwave functions. This new research area, which aims towards a new family of microwave devices, deals mainly with the photoconductivity in semiconductors with 3D, 2D and 1D structures for applications ranging from the generation of microwave signals to the sampling of those signals.

Section 1.2.1 is focused on the identification of the main electromagnetic properties of specific families of matter. Section 1.2.2 discusses the mechanisms of potential optical transitions and light absorption in semiconductors and dielectric or semi-insulating material/metal interfaces. Section 1.2.3 addresses the photoconductivity of semiconductor materials.

When interacting with a material, the electric field E and the magnetic field B that define an electromagnetic wave progressing in time with a pulsation ω and a wave vector k will interact with the electromagnetic properties of the material, such as its permittivity, permeability and conductivity.

During this interaction, the coupling of these fields with the material will generate some effects on the material polarization (see Table 1.2), leading to optical signal wave vector direction modification by wave phase velocity change through optical index tuning

Material type vsPhysical effects	Refraction	Absorption	Reflection/ Diffraction
Dielectric	Electro-optic ($1^{st}/2^{nd}$ order) Acousto-optic		wave mixing
Semiconductor	Electro-absorption	Photoconductor and Photovoltaic effects	Gratings
Metal	Plasmonic	Surface or volume plasmons	Gratings

Table 1.2. *Review of main light/matter interaction*

In this book, the study is limited to two light/matter interactions occurring by absorption such as photoconductive effect and plasmonics, involving semiconducting materials and dielecric/metal interfaces properties, respectively.

1.2.1. *Fundamental electromagnetic properties of 3D bulk materials*

The distinction of semiconducting, dielectric and metallic materials originates from their electronic band structure, i.e. the spatial distribution of electrons and holes inside leading to a conductive or non-conductive behavior. The knowledge of the density of states (DoS) respecting the Fermi golden rule leads to the definition of the density of free electrons in the conduction band and free holes in the valence band. In materials demonstrating a non-zero energy gap between conduction and valence bands (semiconducting materials), highest carrier density values appear at energy levels near valence and conduction band edges such as E_v and E_c. Optical transitions by absorption and emission of energy, which can be radiative or non-radiative processes, can temporally modify these densities by generation of excitons (pairs of free carriers) according to their lifetime.

1.2.1.1. *Electronic band structure (energy band diagram and DoS)*

In 3D bulk materials, if we define the same volume denoted by *Lx, Ly* and *Lz* dimensions in the *x*-, *y*- and *z*-directions, respectively, equivalent to elementary $\frac{\pi}{L_x}.\frac{\pi}{L_y}.\frac{\pi}{L_z}$ volume translated in the *k*-space, then DoS expression is determined by a root mean square profile dependency with energy [ZEG XX].

DoS refers to the number of states per interval of energy at each energy level that are available to be occupied by electrons. In other words, DoS, denoted by *g(E)*, indicates how densely quantum states are packed in a particular system.

Integrating the density of the quantum states over a range of energy will produce a number of states *N(E)*, equation [1.1].

$$N(E) = \int_{E}^{\Delta E} g(E)dE \qquad [1.1]$$

where *g(E)dE* represents the number of states between *E* and *dE*.

From the Schrödinger equation, we know that the energy of a particle is quantized and is given by equation [1.2] where *h* is Planck's constant.

$$E = \frac{k^2\hbar^2}{2m} \qquad [1.2]$$

Bearing in mind that the energy of a particle of mass *m* is given by equation [1.3]

$$E = \frac{1}{2}mv^2 = \frac{m^2v^2}{2m} = \frac{p^2}{2m} \qquad [1.3]$$

We deduce the relation, equation [1.4], between the variable *k* to the physical quantity of momentum *p*.

$$E = \frac{k^2\hbar^2}{2m} = \frac{p^2}{2m} \rightarrow k = \frac{p}{\hbar} \qquad [1.4]$$

The momentum *p* is a vector which has components in the *x*-, *y*-, and *z*-directions.

In a 3D system, the total energy is given by equation [1.5]

$$E = \frac{\hbar^2}{2m}\left(k_x^2 + k_y^2 + k_z^2\right) \qquad [1.5]$$

Knowing that the quantum states are separated by an interval of π/a, the volume of the unit cell (associated with a sphere) is given by equation [1.6]

$$V_0 = \left(\frac{\pi}{a}\right)^3 \qquad [1.6]$$

The DoS problem is a problem of finding the number of states in the interval of E and $E+dE$. In *k*-space, the intervals are simply *k* and *k+dk*. Because we are operating three dimensions, *k* represents the radius of a sphere in *k*-space and *dk* is the thickness of the sphere.

The final result for the DoS is calculated from the expression of *g(E)dE*, the number of states between *E* and *dE* (see equation [1.7]):

$$\begin{aligned} g(E)dE &= \frac{a^3}{\pi^2}\frac{2\left(2^{-1/2}\right)m^2\left(m^{-1/2}\right)E\left(E^{-1/2}\right)}{\hbar^4\hbar^{-1}}dE \\ &= \frac{a^3}{2\pi^2}\frac{2^2\left(2^{-1/2}\right)m^{3/2}}{\hbar^3}\sqrt{E}dE \end{aligned} \qquad [1.7]$$

$$g(E)dE = \frac{a^3}{2\pi^2}\left(\frac{2m}{\hbar^2}\right)^{3/2}\sqrt{E}dE$$

1.2.1.2. *Optical constants and the dielectric function + Kramers–Krönig relationships*

Kramers–Krönig relations describe the connection between the real and imaginary parts of linear complex optical functions descriptive of light–matter interaction phenomena such as susceptibility, dielectric function, refraction index and reflectivity. The real and imaginary parts are connected by a special form of the Hilbert transform. The sum rules are universal constraints that determine the results of integration over the infinite spectral range of the functions descriptive of relevant optical properties of the medium under investigation. By applying the Kramers–Krönig relation, it is possible to acquire knowledge on dispersive phenomena by the measurement of absorptive phenomena on the whole spectrum.

Hilbert transforms connect the real and imaginary parts of a(ω) as follows (equation [1.8]):

$$\mathrm{Re}\{a(\omega)\} = \frac{1}{\pi} \mathrm{P} \int_{-\infty}^{+\infty} \frac{\mathrm{Im}\{a(\omega')\}}{\omega' - \omega} d\omega' \qquad [1.8]$$

$$\mathrm{Im}\{a(\omega)\} = \frac{1}{\pi} \mathrm{P} \int_{-\infty}^{+\infty} \frac{\mathrm{Re}\{a(\omega')\}}{\omega' - \omega} d\omega'$$

Thus, the causality of *a*(*t*), together with its property of being a function belonging to the space of the square-integrable functions L2, implies that its Fourier transform *a*(*ω*) is analytic in the upper complex ω-plane and that the real and imaginary parts of *a*(*ω*) are not independent but are connected by non-local, integral relations called dispersion relations.

The linear susceptibility function describes, at a fundamental level, the connection between the microscopic dynamics of the system under consideration and its linear optical properties. Nevertheless, it is experimentally much easier to measure other quantities that are more directly related to the behavior of light influenced by its interaction with matter. The most commonly used optical constants are the complex index of refraction N (ω) = η(ω) + iκ(ω), and the complex reflectivity at normal incidence r(ω). It has been shown that relevant integral relations can also be established for these quantities.

1.2.1.3. *Free carriers density*

For each type of semiconductor, extrinsic or intrinsic (i.e. doped or non-intentionally doped), at the thermodynamic equilibrium state, the concentrations of electrons n0 (9) and holes p0 (10) are defined according to the DoS in the valence band N_v and in the conduction band N_c. They depend also on the minimum energy in the conduction band Ec, the maximum energy in the valence band Ev and the Fermi level E_F. They can be expressed with respect to the intrinsic concentration n_i of the semiconductor using the action mass low (11), where T represents the temperature in Kelvin and kB represents the Boltzmann constant.

$$n_0 = N_c \exp\left[-\frac{E_c - E_F}{k_B T}\right] \quad [1.9]$$

$$p_0 = N_v \exp\left[-\frac{E_v - E_F}{k_B T}\right] \quad [1.10]$$

$$n_0 p_0 = n_i^2 \quad [1.11]$$

N_c and N_v are calculated from the product of the DoS defined in [1.7] by the Fermi–Dirac distribution function as described in the previous paragraph for 3D materials.

When the material is subjected to an external perturbation such as illumination, these carrier concentrations are modified following equations [1.12] and [1.13] and the law of mass action is no longer valid [1.14].

$$n = n_0 + \Delta n \quad [1.12]$$

$$p = p_0 + \Delta p \quad [1.13]$$

$$np \neq n_i^2 \quad [1.14]$$

The system will try to recover an equilibrium state through multiple recombinations between the free carriers in the photoexcited material. The presence of impurities modifies the structure of the material bandgap. These impurities introduce an intermediate energy level between E_v and E_c, which favors generation-recombination processes at a level called traps level.

Below, three major types of generation-recombination of electron–hole pair processes are identified.

1.2.1.3.1. Thermal (emission or absorption of phonons)

The thermal generation-recombination is linked to the emission or absorption of phonons. These thermal processes are involved in the presence of free carriers in the valence or conduction band in the frame of intrinsic or extrinsic semiconductors under equilibrium state, i.e. without optical pumping or current injection. When the temperature decreases, the thermal generation of carriers decreases; thus, the free carrier concentration diminishes.

1.2.1.3.2. Radiative (emission or absorption of photons)

During the radiative generation-recombination, an electron–hole pair is generated or recombines through the emission or absorption of photons, respectively.

1.2.1.3.3. Auger

The Auger mechanism intervenes when a recombination between an electron and a hole followed by an energy transfer from the recombined pair to a free carrier.

In the case of the refraction of this incoming wave with a wavelength, electro- or acousto-optic interactions will modify the index of the materials and consequently to transmit the incident wave under different angles following the well-known Snell–Descartes law.

The interaction between a light wave and a material will differ according to the nature of the material [WIK]. For a conventional material ($n > 1$), the angle of refraction will be positive, and for an artificial material ($n < 0$), this angle will be negative (see Figure 1.6).

For an interaction with a completely reflecting material, the incident wave will be reflected under an angle equal to the incident wave for a conducting material. If the material is an insulator, the incident wave will be reflected or diffracted under an angle according to the Snell–Descartes law.

This chapter will be dedicated to this kind of material, especially those with a bandgap (semiconductor). A semiconductor material is an insulator that becomes a conductor under an external excitation, which could be thermal or electrical or optical. Under an external excitation, the number of free carriers for each energy band is modified. As an example, an electron in the valence band is transferred into the conduction band leaving a hole in the valence band. The energy brought by the external excitation must be greater than or equal to the gap energy in order to generate the electron–hole pairs.

The photoconductivity or photoconductive effect is based on the phenomenon of photon absorption into a semiconductor material [SAL 07].

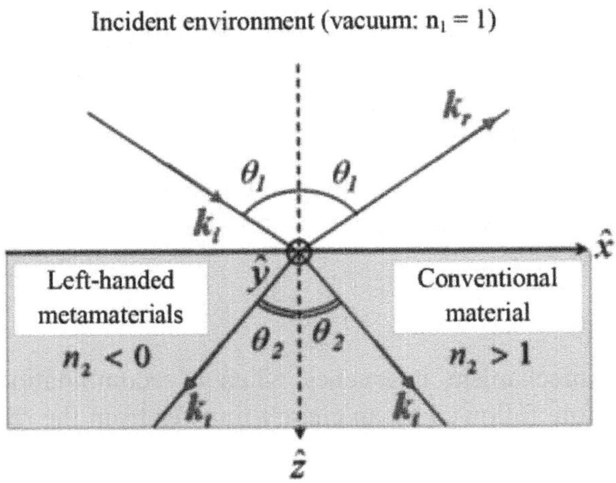

Figure 1.6. *Interaction of a light wave with a conventional material ($n \geq 1$) or an artificial material ($n < 1$). For a color version of this figure, see www.iste.co.uk/tripon/nanotechnology.zip*

Under a light illumination with an energy larger than the bandgap E_g [1.15], free carriers (either electrons or holes) are photogenerated in the material in the illuminated area.

The absorption α of the semiconductor is defined with respect to the wavelength λ of the optical signal associated with the light illumination [1.16]. In these equations, E_g is measured in eV, α is measured in cm^{-1}, h is the Planck's constant and υ is the optical frequency.

$$h\upsilon \geq E_g \qquad [1.15]$$

$$\alpha \approx 4.10^4 \sqrt{(h\upsilon - E_g)} \qquad [1.16]$$

We must therefore consider the optical wavelength/semiconductor couple in favor of the photoconductive effect (see Figure 1.7), which shows absorption coefficient for various semiconductors with respect to the photon energy and the wavelength from 200 nm to 1.4 μm. As an example, in Table 1.3, typical values for α for GaAs and Si are given.

	α (cm^{-1}) λ = 671 nm	α (cm^{-1}) λ = 800 nm
GaAs	2.6×10^4	10^4
Si	4×10^3	10^3

Table 1.3. *Absorption coefficients for GaAs and Si at two different wavelengths*

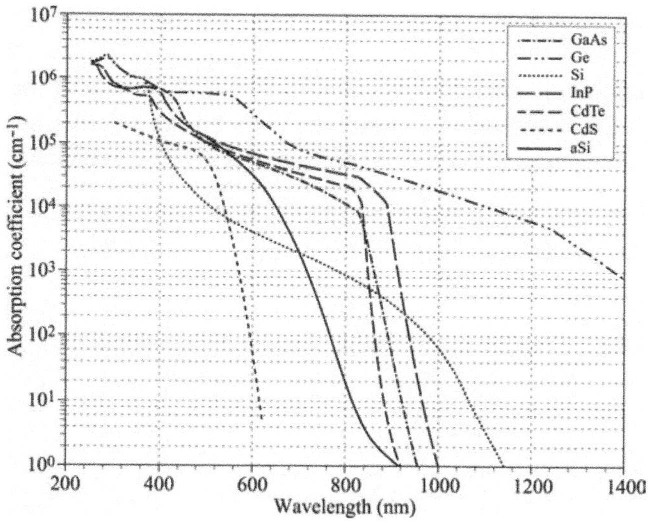

Figure 1.7. *Absorption coefficient for various semiconductors with respect to the photon energy and the wavelength from 200 nm to 1.4 μm*

During the optical illumination, the material absorbs a flux ϕ of incoming photons if the optical wavelength is included in the absorption band of the semiconductor. The expression of this flux can be determined by using the absorption coefficient α per unit of surface and time.

In the direction of the light illumination, the flux follows an exponential low decreasing proportionally to the absorption depth z and the reflection coefficient R of the surface of the semiconductor [1.16]. Its amplitude is maximum at the surface level, i.e. at $z = 0$ (equation [1.17]).

In equation [1.18], P_{opt} represents the incoming optical power and A is the illuminated area of the material, c is the speed of light and q is the charge of the electron.

$$\phi = \phi_0 \, e^{-\alpha z} \quad [1.17]$$

$$\phi_0 = \frac{P_{opt}}{A} \frac{\lambda}{hc} \frac{1}{q} (1 - R) \quad [1.18]$$

In the simplified case of the air/semiconductor interface, the reflectivity or reflectance r [1.19] depends on the material index $\tilde{n}_{sc}$, which in the case of an absorbing material includes an imaginary part representing the absorption [1.20].

The reflection coefficient R is equal to the square norm of the reflectance calculated at the interface [1.21] and depends on the relative permittivity of the semiconductor ε_r and on the absorption at a given wavelength λ.

$$\tilde{n}_{sc} = n - i\kappa = \sqrt{\varepsilon_r} - i\frac{\alpha\lambda}{4\pi} \quad [1.19]$$

$$r = \frac{\tilde{n}_{sc} - 1}{\tilde{n}_{sc} + 1} \quad [1.20]$$

$$R = rr^* = \frac{\left(\sqrt{\varepsilon_r} - 1\right)^2 + \left(\frac{\alpha\lambda}{4\pi}\right)^2}{\left(\sqrt{\varepsilon_r} + 1\right)^2 + \left(\frac{\alpha\lambda}{4\pi}\right)^2} \quad [1.21]$$

1.2.2. *Linear optical transitions*

The optical properties of semiconductors at low light levels are often referred to as linear properties in contrast to the nonlinear optical properties. There are many physical processes that control the amount of absorption or other optical properties of a semiconductor. These processes depend on the wavelength of radiation, the specific properties of the semiconductor being studied, and other external parameters such as pressure and temperature. Optical properties of semiconductors are affected either by the introduction of impurity dopants (both *p* and *n*- type) or by unwanted impurities or defects, similar to electrical properties. Intrinsic optical properties of semiconductors are properties that depend on their perfect crystalline nature, and extrinsic properties are brought by impurities or defects. Many types of defects exist in real solids: point defects, macroscopic structural defects, etc. Figure 1.8 schematically depicts various contributions to the absorption spectrum of a typical semiconductor as functions of wavelength (top axis) and photon energy (bottom axis). Some of the structures shown may be reduced or they are not actually present in a particular semiconductor (e.g. impurity absorption and bound excitons).

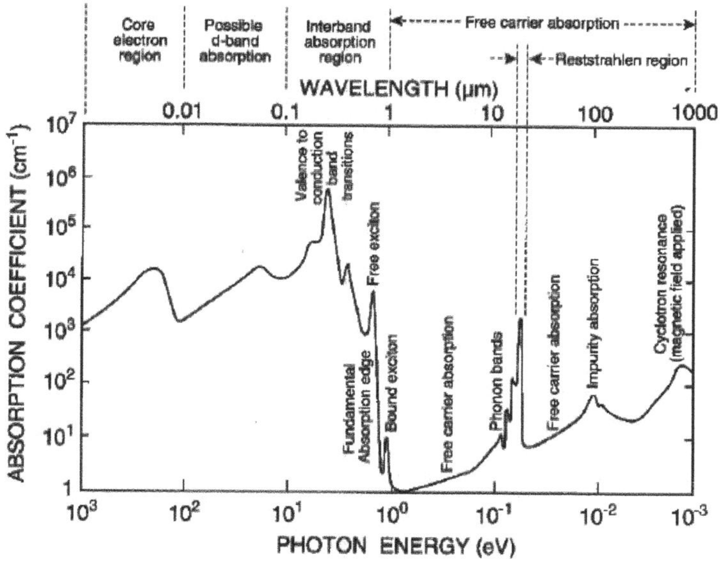

Figure 1.8. *Absorption spectrum of typical semiconductor showing a wide range of optical processes [AMI 95]*

1.2.3. *Bandgap engineering in nanomaterials: effect of confinement/sizing on bandgap structure*

1.2.3.1. *Material particularities of 2D materials: electronic band structure (energy band diagram/DoS) in 2D materials – Van Hove singularities*

In 2D, an electron is confined along one dimension but is able to travel freely in the other two directions. In the image below an electron would be confined in the *z*-direction but would travel freely in the XY plane.

In the 3D DoS analysis, a spherical volume of width *dk* had to be used. However, in 2D, the problem of calculation becomes easier because we only need to operate in two dimensions. In the 2D case, the unit cell is simply a square with a side length of π/a.

The area of the unit cell is given by equation [1.22]:

$$A_0 = \left(\frac{\pi}{a}\right)^2 \qquad [1.22]$$

Because we know the relation between *k* and *E*, we can find DoS as a function of energy.

Substituting the results into the DoS equation will give the DoS in terms of energy [1.23].

$$g(E)dE = \frac{a^2}{\pi}\left(\frac{2mE}{\hbar^2}\right)^{1/2}\left[\frac{m}{\hbar^2}\left(\frac{2mE}{\hbar^2}\right)^{-1/2}\right]dE$$

$$g(E)dE = \frac{a^2 m}{\pi\hbar^2}dE \qquad [1.23]$$

It should be noted that the 2D DoS does not depend on energy.

1.2.3.2. *Material particularities of 1D materials: energy band diagram/DoS in 1D materials*

The DoS for a 1D quantum mechanical system shows a unique solution that has application in materials such as NWs and CNTs. In both *x*- and *y*-directions, the electron is confined, but it moves freely in the *z*-direction.

Going from the 2D case to the 1D case, the unit cell becomes a line segment of length π/a.

Just as a ring was used in the 2D system to find the number of quantum states in a differential interval of energy, so a line segment is used in the 1D situation. The line segment stretches from k to $k + dk$ or E to $E + dE$, and the length of the line is simply dE. Dividing the infinitesimal line length dE by the line length of the unit cell and then multiplying by 2 (i.e. accounting for the two electron spins), the DoS equation is [1.24]

$$g(k)dk = 2\left(\frac{L}{L_0}\right) = (2)\frac{dk}{\left(\frac{\pi}{a}\right)} = \frac{2a}{\pi}dk \qquad [1.24]$$

Replacing dE with dk, the DoS for 1D materials can be written [1.25] as

$$g(k)dk = \frac{2a}{\pi}dk$$

$$g(E)dE = \frac{2a}{\pi}\left[\frac{m}{\hbar^2}\left(\frac{2mE}{\hbar^2}\right)^{-1/2}\right]dE$$

$$= \left(\frac{a}{\pi}\right)\frac{(2)2^{-1/2}(m)m^{-1/2}}{\hbar^2\hbar^{-1}}E^{-1/2}dE \qquad [1.25]$$

$$g(E)dE = \left(\frac{a}{\pi}\sqrt{\frac{2m}{\hbar^2}}\right)\frac{1}{\sqrt{E}}dE$$

A typical representation of the DoS for 1D materials is shown in Figure 1.9.

1.2.3.3. *Material particularities of 0D materials: energy band diagram/DoS in 1D materials*

The DoS for a 0D structure (e.g. QD) is associated with the absence of motion and the absence of k-space to be filled with electrons. As a result, the available states only exist at discrete energies. In that case, the DoS for 0D materials can be expressed as a delta function:

$$g(E) = 2\,(E\text{-}E_c) \qquad [1.26]$$

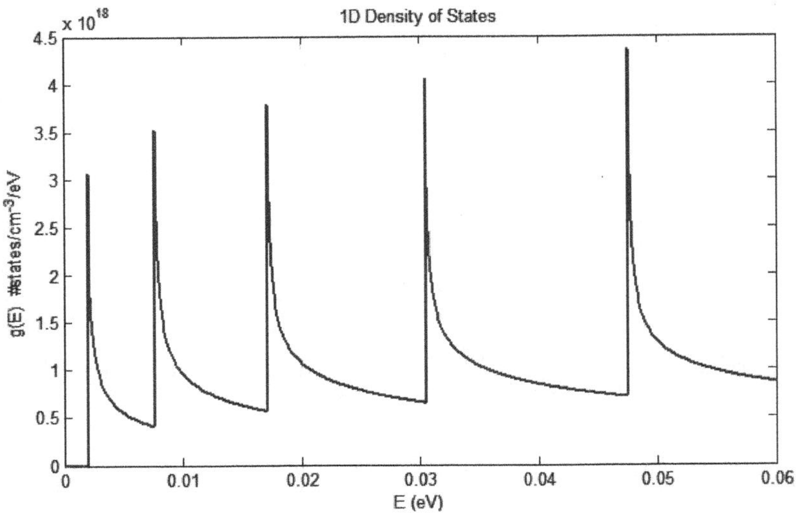

Figure 1.9. *Typical representation of the density of states of 1D materials*

1.3. Focus on two light/matter interactions at the material level

1.3.1. *Photoconductivity in semiconductor material*

1.3.1.1. *General description and some elements on the history of the field*

As an introduction, some major pioneering contributions in the field of photoconductivity have been selected.

First, we recall the work presented in 1873 by Willoughby Smith [SMI 73] who discovered photoconductivity in selenium while testing the metal for its high resistance properties in conjunction with his work involving submarine telegraph cables.

Since this early discovery, a strong effort has been made to control the speed and the efficiency of this new effect found in semiconductors

This pioneering work was followed later by a major paper by Albert Rose in 1955 [ROS 55], which concentrated on the performance of photoconductors, where he stated that "all semi-conductors and insulators are photoconductors that the characteristic parameter of a photoconductor is the lifetime of a free carrier". The photoelectron current is equal to the product of absorbed photon current and the ratio of lifetime to transit time of a free carrier. This relation holds as well for the commonly known types of barriers as for uniform photoconductors. The photoelectron current may range from a small fraction of the photon current to many powers of 10 greater than the photon current depending on the ratio of lifetime to transit time. There is good evidence for lifetimes in different photoconductors extending from 10^{-12} s to values approaching a second. The lifetimes of free electrons and free holes are, in general, independent of each other and significantly different. Only at sufficiently high excitation rates for which the free carrier densities exceed the densities of bound states do the electron and hole lifetimes necessarily become equal. Some decades later, the time response of a photoconductor was deeply analyzed by B. Winter and E. Rosencher [ROS 98].

1.3.1.2. *Material photoconductivity*

The absorption of energy in these materials will generate localized charge carriers electron density Δn in the valence band and generation of holes density through a transfer toward the conduction band owing to their displacement in the material and the generation of a current if the material is submitted to an electric field. In this configuration, this increase of free carriers depends directly on the duration of the optical illumination and on the lifetime $\tau_{n,p}$ of the electrons/holes in the conduction/valence band before their recombination in order to reach an equilibrium state.

The intrinsic conductivity of a semiconductor material can be expressed directly as a function of the carrier density at the thermal equilibrium through equation [1.27]:

$$\sigma_0 = q\left(n_0\mu_n + p_0\mu_p\right) \qquad [1.27]$$

The photoconductivity resulting from the increase of the carrier density of the semiconductor through photon absorption is written as equations [1.28] and [1.29]:

$$\sigma = \sigma_0 + \Delta\sigma(x) \qquad [1.28]$$

$$\Delta\sigma(x) = q\left[\mu_n \Delta n(x) + \mu_p \Delta p(x)\right] \qquad [1.29]$$

The material photoconductivity is defined from the increase of the free carrier concentrations $\Delta n(x)$ and $\Delta p(x)$ and from the conductivity at equilibrium σ_0.

If we consider a semiconductor material located between two electrodes a length L apart and submitted to an optical illumination orthogonal to the surface length L and width W with a thickness D, the photogenerated current between these two electrodes is defined by equation [1.30] as a function of the electric field E present in the semiconductor material and the drift velocity of the electrons v_d

$$I_p = \sigma E.W.D = q\mu_n nE.W.D = qnv_d.W.D \qquad [1.30]$$

As the first approximation, the generation rate of free carriers G can be written by equation [1.31], introducing the quantum efficiency η (number of generated electrons per photons), the injected optical power P_{opt} at the optical frequency ν:

$$G = \frac{\Delta n}{\tau} = \frac{\eta.P_{opt}/h\upsilon}{W.L.D} \qquad [1.31]$$

The photocurrent can be defined by equation [1.32] as a function of the primary photocurrent I_{ph} in the case where the carrier lifetime is similar to the transit time of the electrons between the electrodes:

$$I_p = q.\left(\eta.\frac{P_{opt}}{h\upsilon}\right).\left(\frac{\mu_n\tau.E}{L}\right) = I_{ph}.\left(\frac{\mu_n\tau.E}{L}\right)\text{with}I_{ph} = q.\left(\eta.\frac{P_{opt}}{h\upsilon}\right) \qquad [1.32]$$

The photoconductive gain is written [1.33] as the ratio of the lifetime τ and the transit time of the electrons:

$$G = \frac{\tau_t}{\tau} = \frac{\eta.P_{opt}/h\upsilon}{W.L.D} \qquad [1.33]$$

The work done by Platte and B. Sauerer allowed us to approximate this spatial evolution effectively to a great extent. This equation permits us to approximate the increase of photoconductivity under optical excitation as a function of the material absorption α, the surface recombination speed of the photogenerated carriers v_s, their lifetime τ and their ambipolar diffusion length L [1.34–36]:

$$\Delta\sigma_m = \frac{\Delta\sigma_{ph}}{1+\alpha L}\left[\frac{1}{\alpha L}\left(\frac{\alpha L^2 + v_s\tau}{L + v_s\tau}\right)\right]^{\frac{-\alpha L}{1-\alpha L}} \qquad [1.34]$$

$$\text{with } d_e = \frac{1}{\alpha}\left[\frac{L(1+\alpha L)}{L+v_s\tau}\right]\left[\frac{1}{\alpha L}\left(\frac{\alpha L^2 + v_s\tau}{L + v_s\tau}\right)\right]^{\frac{\alpha L}{1-\alpha L}} \qquad [1.35]$$

$$L = \sqrt{\frac{2k_B T}{e}\tau\left(\frac{\mu_n \mu_p}{\mu_n + \mu_p}\right)} \qquad [1.36]$$

As a result, the direct representation of the photoconductive effect is a resistance R_g inversely proportional to the increase of the photogenerated conductivity G_g.

While integrating equation [1.36], it is possible to write the photoconductance G_g by considering that the current flows through a plane (yOx) following the discontinuity length L_g, i.e. along the Oz axis.

The variation of the photoconductance G_g of the photos, which is written by equation [1.37], becomes an integral along $\vec{x}$ (xx) with a uniform illumination in the (yOz) plane

$$G_g = \int \Delta\sigma(x)\frac{dS}{dz} = \int \Delta\sigma(x)\frac{dydx}{dz} \qquad [1.37]$$

$$G_g = \frac{w_{eff}}{L_g}\int \Delta\sigma(x)dx \qquad [1.38]$$

So when the semiconductor gap is illuminated by an optical beam at a wavelength λ, with a constant optical power P_{opt}, and a diameter w_{eff}, a photoconductance G_g (equations [1.39] and [1.40]) is generated at the level of the illuminated area A, in parallel to the capacity C_g, with a quantum efficiency η (close to 1), where R is the reflection coefficient of the substrate:

$$G_g = \frac{w_{eff}\Delta\sigma_{ph}}{L_g\left(1-\alpha^2L^2\right)}\left(\frac{1}{\alpha} - L\frac{\alpha L^2 + v_s\tau}{L + v_s\tau}\right) \qquad [1.39]$$

with $$\Delta\sigma_{ph} = \frac{e\lambda}{hc}\frac{P_{opt}}{A}\left(\mu_n + \mu_p\right)\eta\alpha\tau(1-R) \qquad [1.40]$$

$$R_g = \frac{1}{G_g} = \frac{L_g\left(1-\alpha^2L^2\right)}{w_{eff}\Delta\sigma_{ph}\left(\frac{1}{\alpha} - L\frac{\alpha L^2 + v_s\tau}{L + v_s\tau}\right)} \qquad [1.41]$$

The constant illumination of a semiconductor material has been simulated by a resistance R_g defined by G_g equation [1.41]. This equation allows us to find the value of the generated photoresistance as a function of the intrinsic parameters of the substrate and of the illumination.

1.3.1.3. *Two-photon absorption*

Some semiconductor materials can simultaneously absorb two photons' energy equal to half the bandgap energy E_g.

This behavior is associated with materials, which have complex index κ equations [1.42] and [1.43], with a nonlinear imaginary part proportional to the light intensity I.

These materials are characterized by an two-photon absorption coefficient $\beta(\omega)$ which depends on the optical pulsation ω, on the index n, on the bandgap energy E_g of the material equation [1.44] and on the Kane's momentum P equivalent to equation [1.31] for most of the semiconductor materials.

For GaAs, the value of the coefficient $\beta(\omega)$ is equal to 5 cm/GW at λ = 1.55 µm at room temperature (T = 300 K). For LT-GaAs, $\beta(\omega)$ reaches 35 cm/GW.

$$\kappa = \kappa_0 + \beta(\omega)\frac{c}{2\pi\upsilon}I \qquad [1.42]$$

with $\beta(\omega) = 724\frac{\pi}{5}\left(\frac{e}{hc}\right)^2 \frac{\hbar P}{n^2 E_g^3} \frac{\left(2\hbar\omega / E_g - 1\right)^{\frac{3}{2}}}{\left(2\hbar\omega / E_g\right)^5}$

$$P = \hbar\sqrt{\frac{10,5}{m_0}} \qquad [1.43]$$

$$\kappa_0 = \frac{\alpha\lambda}{4\pi} \qquad [1.44]$$

These materials can be optically excited at an optical wavelength λ_{2p} equal to twice the absorption wavelength λ_g corresponding to E_g. The first photon will allow an electron in the valence band to reach a superior intermediate energy level centered in the bandgap. The second photon transfers its energy to this electron, which will have access to the conduction band.

This nonlinear absorption effect in these materials opens the way to the realization of optically controlled circuit materials with a high-power optical source at a wavelength equal to $2\lambda_g$.

As an example, for GaAs, it could be interesting to use an optical source at 1.55 µm corresponding to the conventional telecommunication wavelength.

1.3.1.4. *Recent 1.55 µm semiconductor materials*

Conventional ultrahigh speed photoconductors are based on semiconductor materials such as LT GaAs sensitive to the wavelength of 0.85 µm. Widely known impulse laser diodes have been developed at 1.55 µm with high repetition rates in the few GHz range for fiber optic telecommunications. However, at 1.55 µm, the LT GaAs is no longer sensitive, and it was necessary to study semiconductor material with characteristics equivalent to LT GaAs in terms of dark resistivity and lifetime or recombination time but operating at 1.55 µm.

At the beginning of the 21st Century, numerous research activities have been developed to realize materials at 1.55 µm, such as indium gallium arsenide (InGaAs) and LT InGaAs, [MAN 07] together with the insertion of ErAs nanoparticles in InGaAs [ZHA 10].

Now the research activities are focused on:

1) ternary compounds such as GaBiAs [PAC 09];

2) quaternary compounds such as GaAsSbN [TAN 13] and GaInAsN [GRA 12];

3) two-photon absorption in GaAs photonic crystals [COM 08];

4) Fe-doped InGaAs [HAT 11];

5) Fe-doped InGaAsP [FEK 11];

6) InGaAs/InAlAs [SAR 08].

Table 1.4 presents some elements for the comparison of performances of new ultrahigh speed photoconductor operating in the 1.55 µm region.

Type of semiconductors	Response time (ps)	Dark resistivity (Ohm.cm)	Mobility (cm^2/V.s)
GaAsBi	1	60	2,000
GaInAsN	4.4		
GaAsSbN	1.3	1×10^7	
TPA GaAs	1		
LT-InGaAs	0.2	3	490
Fe doped InGaAs	0.6	2,200	
ErAs/InGaAs superlattice	0.2 – 0,6	340	
InGaAs/InAlAs	0.75	10^6	
Be-doped InGaAs/InAlAs multilayer	1	100–200	500– 1,500
Cold Fe-implanted InGaAsP	0.3–3	1,200 Up to 2,500	400

Table 1.4. *Comparison of performances of new photoconductor operating at 1.55 µm*

1.3.1.5. *Interest for increasing the confinement and low-dimensional materials*

Recent research involving UPMC (Paris 6 University), IEMN (CNRS and University of Lille) and Thales have shown the importance of reducing the gap dimensions in order to increase the photoconductance. As shown in Figure 1.10, when reducing the gap length Lg and the optical beam diameter, the photoconductance increases exponentially. Numerical models of photoconductivity through equation [1.26] help us to determine potential high photoconductance as a function of optical beam diameter, gap length and width dimension tunings. Also in Figure 1.10, the photoconductance Gg value is calculated as a function of optical beam diameter for different gap lengths. As a result, a reduction of gap length from 50 to 0.5 m improves photoconductance by a factor of 100 with an optical beam diameter of 62.5 µm (multimode fiber – MMF). Also, an optical beam diameter of 5 µm (single-mode fiber – SMF) enhances Gg by a factor of 10 for a gap length of 0.5 m compared with the same power illumination with an optical beam diameter of 62.5 µm. Then, an optical focusing system allows us to reduce the gap length and increase the On/Off ratio [TRI 12].

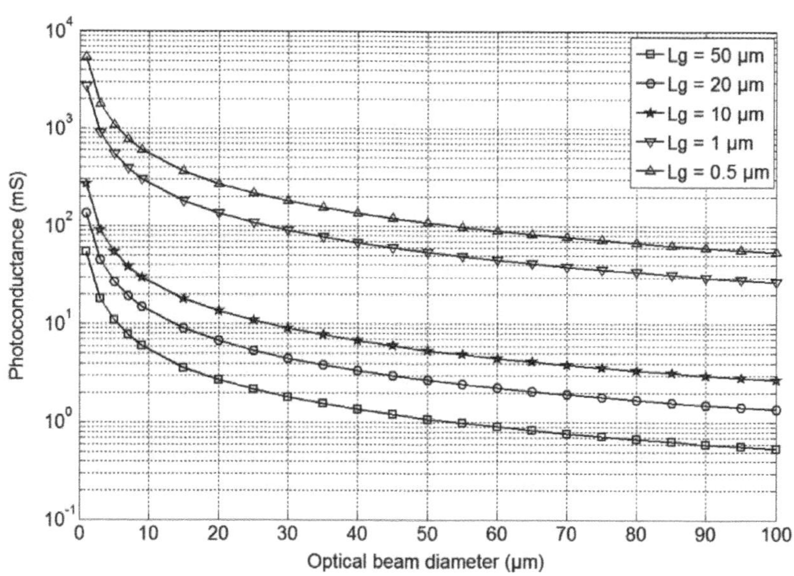

Figure 1.10. *Photoconductance as a function of optical beam diameter for various gap lengths (GaAs substrate, $\lambda = 0.8\ \mu m$, $P_{opt} = 100\ mW$, $\eta = 1$, $P = 30\%$, $\mu_n + \mu_p = 8900 cm^2/Vs$, $\alpha = 4.10^4 cm^{-1}$ $v_s = 10^4 cm/s$, $L = 2\mu m$, $\tau = 200\ ps$)*

1.3.1.6. *General optical properties of SW CNTs (1D)*

CNT-based optoelectronics devices started their emergence 10 years ago because of their exceptional material optical properties in terms of band diagram, delivering direct energy band transitions and drastically reducing phonon-assisted transitions responsible for noise in global performances and low defect density, compared with classical bulk semiconducting materials. In this way, photoconductive and photovoltaic effects in SW and MW CNTs are mainly exploited to develop NIR photodetectors. Today, this material acts as a new contributor for THz signal generation by photomixing.

1.3.1.6.1. Electronic nature of SWCNTs

The electronic properties of CNT are defined from a graphene monolayer where carbon atoms are spatially distributed as a periodic honeycomb geometry (Figure 1.11). From the growth process, CNTs are formed as an axial graphene sheet rolling following a chiral vector Ct relying on two arbitrary vectors a_1 and a_2 separated by an angle ϑ as

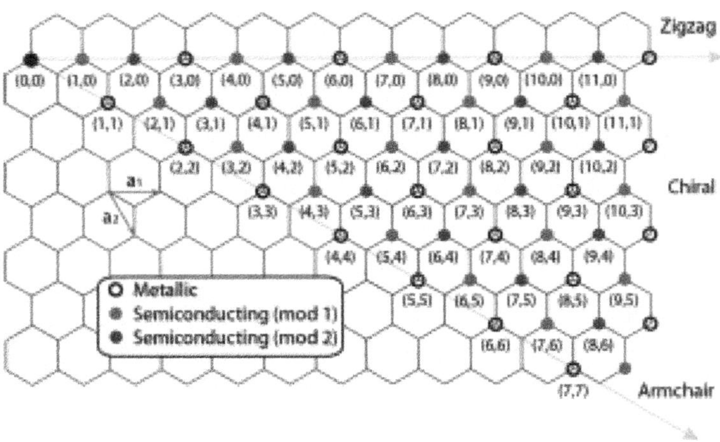

Figure 1.11. *Representation of CNT electronic nature with chirality indices (n,m). For a color version of this figure, see www.iste.co.uk/tripon/nanotechnology.zip*

$$\vec{C}_h = n\vec{a}_1 + m\vec{a}_2 \quad [1.45]$$

$$\cos\theta = \frac{2n+m}{2\sqrt{n^2+nm+m^2}} \quad [1.46]$$

The two n and m integers define the chirality of a CNT by imposing the direction of C_h, which can be identified as zigzag, armchair or strongly chiral.

This (n,m) integer pair also determines the CNT electronic nature as a semiconducting or metallic nanomaterial if $m = 3N$ and $n = 0$, $m = n$, $2m + n = 3N$, in a zigzag, armchair or chiral configuration, respectively, as in equations [1.45] and [1.46].

CNT diameter is also related to this (n,m) integer pair as expressed in [1.47].

$$d_t = \frac{\left|\vec{C}_h\right|}{\pi} = \frac{a_0}{\pi}\sqrt{n^2 + nm + m^2}$$

with $a_0 = \sqrt{3}.a_{C-C} = 0{,}246\ nm$ [1.47]

1.3.1.6.2. SW/MW CNT energy band diagrams

Optical absorption and emission properties of SW/MW CNTs rely on their energy band diagram characterized by unconventional conduction and valence bands profiles with discretized energy spikes values known as Van Hove singularity pairs (Figure 1.12). The energy separation between the highest valence band and the lowest conduction band singularities is given by [1.48] and [1.49], where $a_{c\text{-}c}$ is the lattice distance of two carbon atoms (~0.142 nm) and d_t is the SW CNT diameter.

$$E_{11}^{s}(d) = 2a_{c-c}\gamma_0/d_t \quad [1.48]$$

$$E_{11}^{m}(d) = 6a_{c-c}\gamma_0/d_t \quad [1.49]$$

As shown in Figure 1.12, the first Van Hove peaks determine the authorized transfer energy for both semiconductor and metallic SW CNT [MIN 01], from a global energy band diagram study. Thus, optical transitions occur at $E_{11}^{s}(d), 2E_{11}^{s}(d), E_{11}^{m}(d), 4E_{11}^{s}(d)$ .... energy levels, which are determined experimentally from fluorescence and emission spectra from photoluminescence or electroluminescence and/or absorption spectra from Raman spectroscopy. These energy levels strongly depend on the tube diameter and chirality vector.

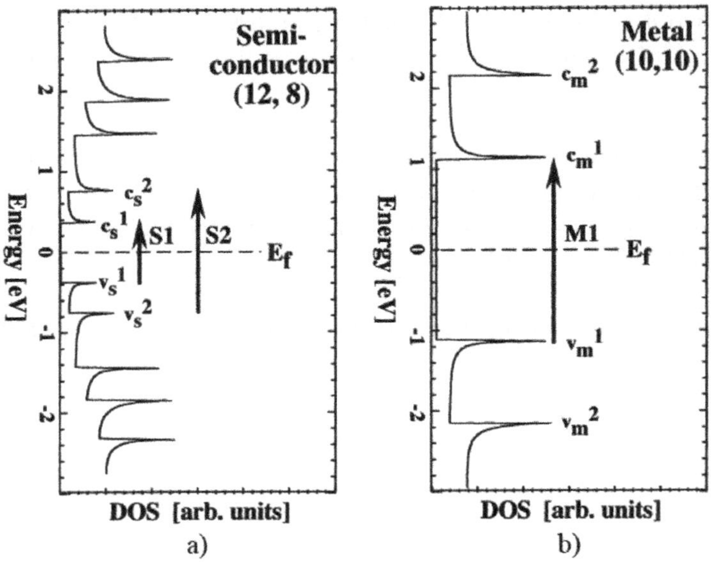

Figure 1.12. *Energy levels as a function of DoS for a) semiconducting and b) metallic SW CNT[PIE 101]*

1.3.1.6.3. Semiconducting and metallic SW CNT optical absorbance spectra

From a classical description of electron/hole pair propagating on cylindrical surface, absorption spectra can be modeled by [1.50] [PED 04]:

$$\alpha(\omega) \propto \frac{\Sigma|\psi_n(x,y)|^2}{(E_n - E_g).(E_n + E_g - \hbar\omega)^2 + (\hbar\Gamma)^2} \qquad [1.50]$$

where $\psi_n(x,y)$ is the *n*th exciton state with energy E_n, $\hbar\omega$ is the photon energy of light excitation, $\hbar\Gamma$ is the phenomenological line width of 0.05 eV, and $E_g = E_{exc} - E_{bind}$ is the bandgap of the CNT, where E_{bind} is the exciton binding energy and E_{exc} is the exciton energy. From this formula, absorption spectra appear as a sum of Lorentzian curves depending on the diameter of the nanotube through the E_g parameter, covering a wideband optical wavelength spectrum from UV to NIR.

For semiconducting SW CNT, two typical absorption peaks characterizing the lowest and the second interband electronic transition, referred to as S_1 and S_2 in Figure 1.13, are well identified with an energy of 0.7 eV and 1.2 eV, respectively.

Absorbance spectra can be shifted by doping, high-pressure treatment or chemical exposure [WAN 12, ZHA 06]. Also, in the case of a multilayered CNT film characterized by a specific thickness and an average nanotube diameter, absorbance optical spectra can be modified by an S1 peak shift of few meV.

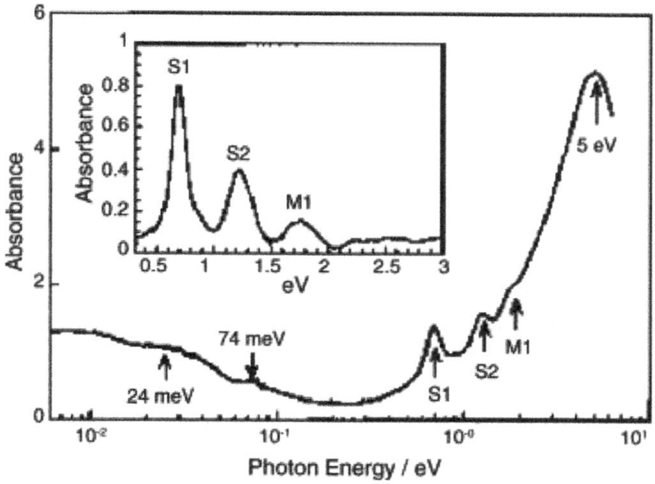

Figure 1.13. *Optical absorption spectrum of SWNT film in air at atmospheric pressure in a wide range of energy from UV to far-IR [PIE 10]*

In the framework of this project, absorbance spectra of commercial CNT are identified by the provider NanoIntegris for both semiconducting and metallic SW CNT, as shown in Figure 1.14. An absorbance of 50% at a wavelength of 0.8 μm is expected for a CNT mixture of 99% semiconducting and 1% metallic.

1.3.1.6.4. Semiconducting absorption coefficient and quantum efficiency

Absorption peaks (Van Hove singularities S11 and S22) of typical CNT types with 1-nm diameter are fairly well matched with commonly used infrared spectrum. An equivalent absorption coefficient of $24 \times 10^4 cm^{-1}$ is reported for CNT bundles, which is high compared to the absorption coefficient of LT GaAs (6,000 cm^{-1}). This point shows another potential advantage of CNTs as compared to semiconducting material [HAQ 06].

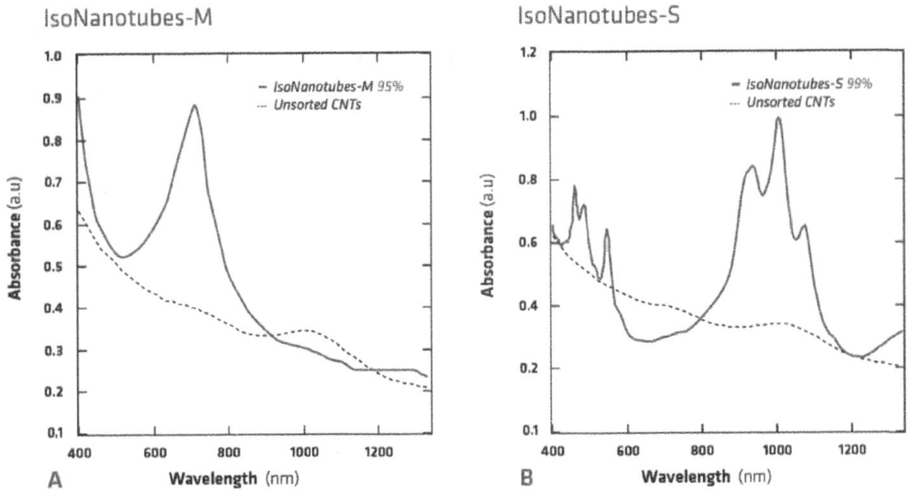

Figure 1.14. *a) Absorbance spectra of metallic and b) semiconducting SW CNT delivered by NanoIntegris*

External quantum efficiency η_e representing the percentage of photons that are able to excite electron–hole pairs depends on both the material and the test device structure. In the case of optical THz photomixing, the 10% quantum efficiency value corresponding to CNTs is lower than the 40% quantum efficiency value achieved in LT GaAs under the same experimental conditions [HES 02].

Energy transitions are dependent on tube diameter, implying an absorption coefficient wavelength dependency. Specific optical measurements such as Raman spectroscopy are necessary to check the spectral absorption band of the material (Figure 1.15).

1.3.1.6.5. Light excitation polarization dependency

Considering p and p' integer scripts for π valence and π* conduction bands, optical transitions occur under parallel or perpendicular light polarization for a p-p' difference equal to 0 or ±1, respectively. In the last case, due to depolarization effects, transitions are suppressed leading to allowed optical transitions only under parallel polarized light. This effect has been characterized by electroluminescence in [KIN 10] and by photocurrent measurement in [QIU 05] (Figure 1.16).

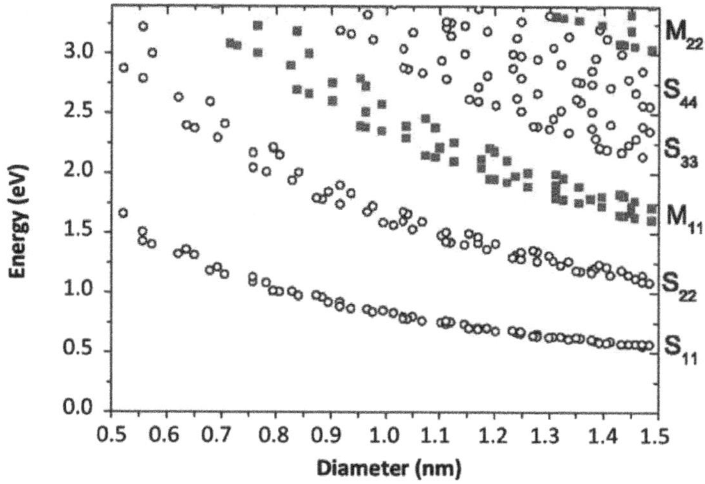

Figure 1.15. *Energy transition versus SWCNT diameter [DRE 05]*

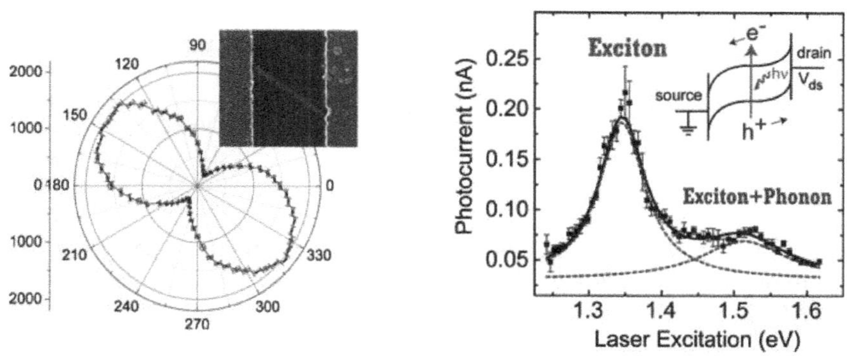

a) b)

Figure 1.16. *Polarization and wavelength dependence of the photocurrent. a) Polarization dependence of the photocurrent. The photocurrent is maximized for linearly polarized light along the length of the CNT. b) Wavelength dependence of the photocurrent. The photocurrent peaks at the expected S22 exciton transition [QIU 05]*

1.3.1.7. *Metal/CNT contact-type influence*

For optoelectronic applications, metal characteristics of CNT-based devices must be studied for electrical characterization and understanding of electron and holes-generated photocurrent behavior, as in many cases a Schottky barrier is created at the CNT/metal interface. The work function of

CNT estimated to 5 eV determining this barrier height with considered metal justifies the choice of asymmetrical electrodes' access in order to avoid carrier photocurrent suppression [HON 10a, Figure 1.17]).

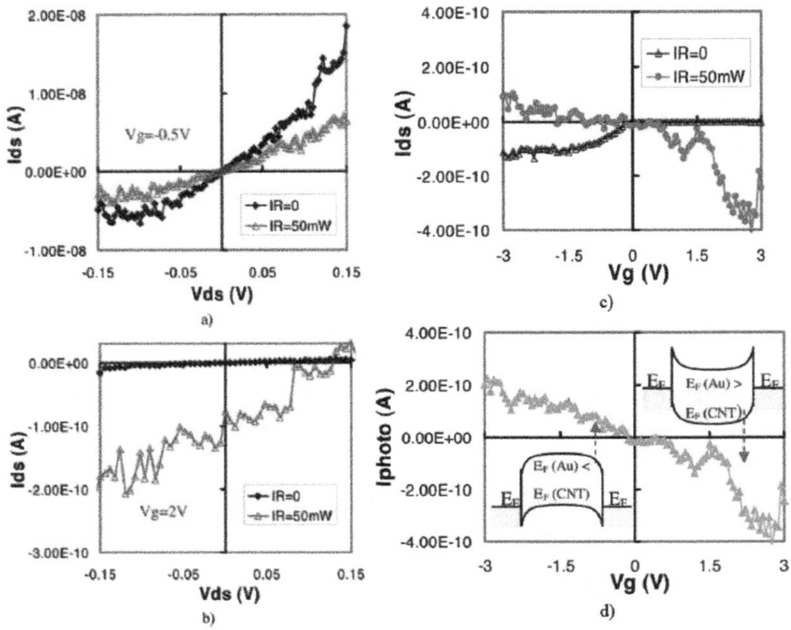

Figure 1.17. *a) Transfer characteristics of the Au-CNT-Au CNTFET with IR source OFF (IR = 0) and ON (IR = 50 mW) at zero bias. b) Relationship between gate voltage and photocurrent and corresponding band diagrams at negative and positive gate voltages – bias-dependent measurement of the Au-CNT-Au CNTFET with IR = 0 and IR = 50 mW at c) Vg = −0.5 V and d) Vg = 2 V*

1.3.1.8. *DC optoelectronic properties*

The photoconductive behavior of CNTs mainly depends on three criteria:

1) CNT orientation to electrical accesses, which enables photocurrent measurement under bias or in a bias-free configuration. For this criterion, CNT deposition method will;

2) CNT density, which also determines the contact resistance value with electrodes' metallization;

3) CNT contact type with electrical accesses metal which can rapidly establish a different Schottky barrier for electrons and holes leading to a potential photocurrent annealing as in CNT-based infrared photodetection [HON 10b].

From DC electrical characterization, the key parameter extraction such as photoconductance can be achieved by photocurrent measurement, assuming the active area volume is determined in two identified configurations as follows:

– Planar unique or bundles of CNT: in this case, the active area volume V is estimated by the product of the surface of a CNT (L_{CNT} x $2\pi d_{CNT}/2$) with a thickness equal to an atomic layer and with the number of CNTs;

– Vertical CNT bundles (N × M CNTs): in this case, the active area volume V is estimated by the product of the active surface of a CNT (de × 2πdCNT/2) with a thickness equal to an atomic layer and with N × M.

1.3.1.8.1. State of the art for horizontal unique SW CNT case

In the case of CNTs deposited in the same plane as electrical access, the benefit of CNT-based electronic devices such as p-n junctions and FETs electrical characterization under specific illumination is crucial.

In the case of p-n junctions, illumination absorption can be transferred on a unique CNT through two processes: first, the excitation of one electron from valence band to conduction band by the absorption of a photon with energy larger than the bandgap and, second, a photon-assisted tunneling process when the photon energy is less than the bandgap. Calculation of the generated photocurrent as a function of the incoming photon energy leads to a spectral photoresponse characterized by multiple peak values due to the multiple direct bandgaps.

For both types of CNT-based electronic devices, I–V characteristics have demonstrated a change in current under bias on one (p-n junction) or two contacts (FET) associated with IR illumination, as shown in Figure 1.18 from [FRE 03] considering a CNT-FET and in Figure 1.19 from [GUO 06] considering metallic and semiconducting CNT-SBFET.

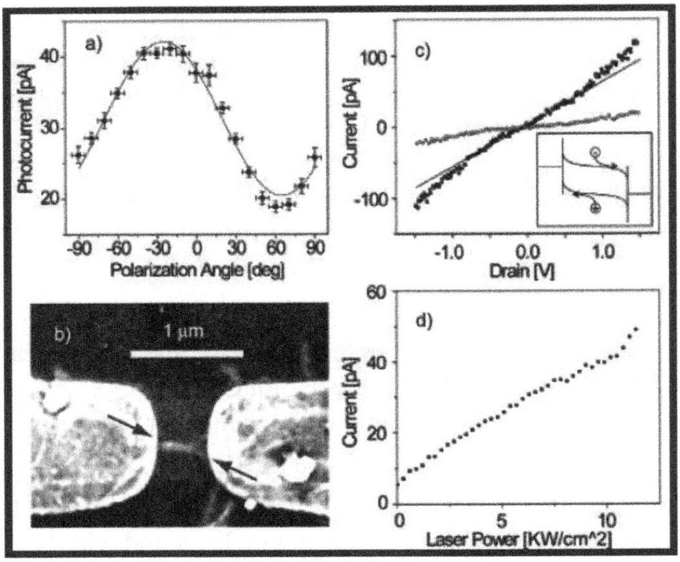

Figure 1.18. *a) Photocurrent as a function of the polarization angle of the incident infrared light. (Excitation energy 1.28 eV, drain voltage +1 V and gate voltage −1.5 V.) b) SEM image of the NTFET. The maximum current in c) is observed with light polarized along the nanotube axis (indicated by the arrows). d) Drain-voltage dependence of the current (O) without light and with infrared light [GUO 06]. (Gate voltage, −2 V; IR intensity, 5.6 KW/cm²) [FRE 03]*

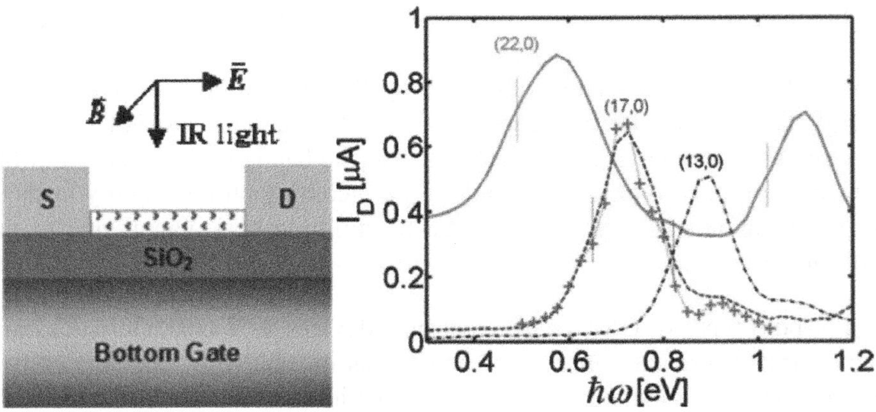

Figure 1.19. *The source-drain current versus the photon energy for a (22,0) CNT with $E_g \approx 0.49$ eV (the solid line), a (17,0) CNT with $E_g \approx 0.63$ eV (dashed line) and a (13,0) CNT with $E_g \approx 0.82$ eV (dashed-dotted line) under the illumination intensity of 107 W/cm² at VG = VD/2 = 0.2 V in the presence of electron–phonon coupling. For comparison, the dotted line with crosses shows the current of the (17,0) CNT without electron–phonon coupling. The vertical bars show the subbandgap*

1.3.1.8.2. State of the art for SW CNT film case

For a film constituted of planar CNTs array, photocurrent measurements have been performed by using a two-electrode device underneath the film allowing bias of the structure [LIU 09]. In this example, the photoconductivity showed an increase, decrease or even negative values when the laser spot was on different positions between contact electrodes, showing a position-dependent photoconductivity of partially aligned films of CNTs. Photon-induced charge carrier generation in SWCNTs and subsequent charge separation across the metal–CNT contacts are believed to cause the photoconductivity changes.

A net photovoltage of 4 mV and a photocurrent of 10 μA were produced under the laser intensity of 27 mW with a quantum efficiency of 7.8% in vacuum (Figure 1.20). The photocurrent was observed to be in the direction of nanotube alignment. Finally, it was found that a strong dependence of the polarization of the incident light on the photocurrent and the orientation of the films influenced the dynamics of the rise and fall of the photocurrent.

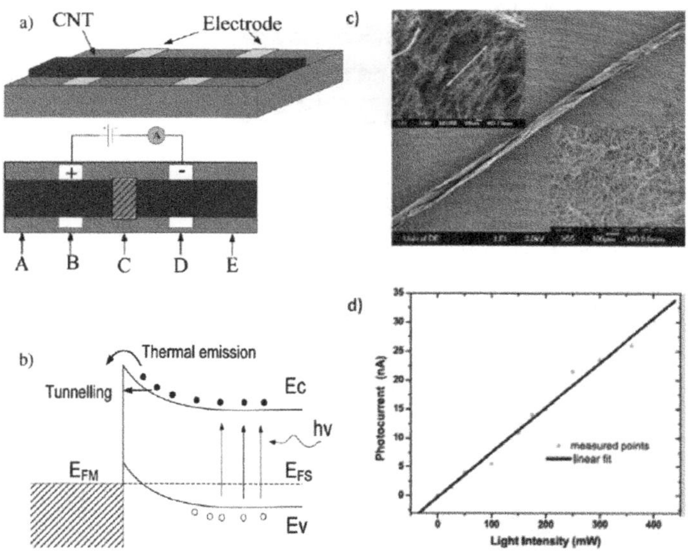

Figure 1.20. *a) Schematic of the carbon nanotube sheets under testing. The bottom part shows the top view of the device. A–E are five testing positions on the sample where a laser spot was pointed. The shaded area indicates that the laser spot was on position C. b) Band diagram of carbon nanotube (right) in contact with platinum electrodes (left). c) SEM image of a partially rolled thin film of carbon nanotubes; inset compares the non-oriented (bottom) and partially oriented (top) samples. d) The amplitude of photocurrent as a function of laser intensity. Measurements were done in an open environment*

1.3.1.8.3. State of the art for MW CNT bundle case

Through experimental comparison, it was found that macroscale CNT materials also have potential distinct optical properties. Photoresponse in bundles of MWNTs was observed by Wei *et al.* [WEI 06], Passacantando *et al.* [PAS 08] and Coscia *et al.* [COS 09]. A recent study has also shown that a MWNT film is capable of generating a position-dependent photocurrent [SUN 08, STO 08]. Photoresponse of MWNT/polymer composite films [STO 08] and photoconductivity of doped nanotubes [JIJ 06] have been studied. There were very few studies about vertically aligned MWNTs before [ZHA 10]. In this last study, length dependency on detected photocurrent level was demonstrated, leading to an estimation of the penetration depth of light larger than 300 μm in this orientation (Figure 1.21).

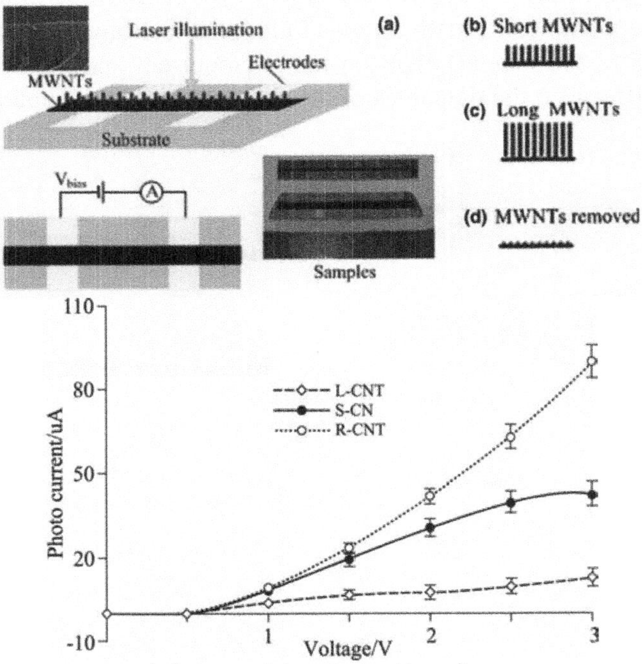

Figure 1.21. *a) Schematic of the MWNTs device under testing. The bottom part shows the top view of the device. The electrodes are Mo; the applied voltage on the electrodes is the bias voltage, Vbias; b) short-aligned MWNTs device (20 μm length); c) long-aligned MWNTs (300 μm length); d) MWNTs removed device; e) photocurrent dependent on bias voltage at room temperature, setting laser power of 27 mW and wavelength at 532 nm. L-CNT means long-aligned MWNTs device, S-CNT means short-aligned MWNTs, R-CNT means MWNTs removed device*

1.3.1.8.4. Carrier dynamics extraction

The key parameters of carriers such as lifetime and mobility are determinant of the estimation of transient photoconductivity σ_{ph} calculated from the sum of electrons and holes conductivity σ_n from [1.51] and [1.52], considering the photogenerated electron *n* and the hole density p:

$$\sigma_n = qen\mu_n \quad [1.51]$$

$$\sigma_p = qen\mu_p \quad [1.52]$$

Photoconductance value is calculated by integration of σ_{ph} over a time corresponding to the carrier lifetime in an active volume.

A real-time investigation of ultra-fast carrier dynamics in SWCNT bundles using femtosecond time-resolved photoelectron spectroscopy has been developed and applied to identify carrier lifetimes value in CNTs bundles [HER 02]. A pump–probe scheme allows an investigation of charge carriers from 2.38 eV above E_F to the immediate vicinity of the Fermi level and below, down to about −0.2 eV, and the study of fundamental scattering processes such as e–e or e–ph scattering directly in the time domain (Figure 1.21).

A carrier lifetime of 1–3 ps with a mobility of 10^4 $cm^2/V.s$ has been obtained experimentally [HES 02] (Figure 1.22).

1.3.2. *Example of light absorption in metals: plasmonics*

1.3.2.1. *Elements on the index and phase conditions for a conventional metal/dielectric interface*

We consider the generation of surface plasmons (oscillation of electrons at a frequency defined as the plasma frequency) at the metal/dielectric interface. These surface plasmons are excited by an electric field incoming in the incident plane of the excitation (typically an optical excitation in tranverse electro magnetic (TEM) mode) and following a specific angle.

Under these conditions, the light wave incoming on the metal/dielectric interface (Figure 1.23) will be partially or totally absorbed. In this configuration, no reflection of this light wave is obtained. The reflection spectrum at the metal/dielectric interface as a function of the incidence angle will reach a minimum value at an angle of incidence called SPR angle, θ_c.

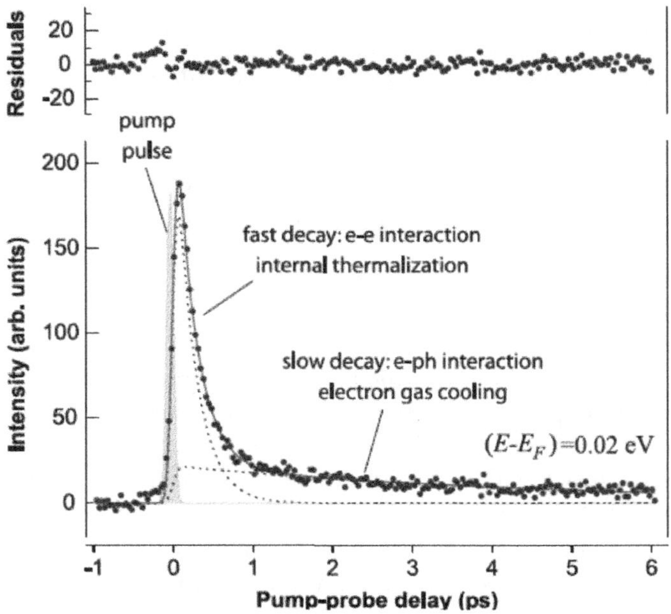

Figure 1.22. *Sample fit of a low-energy cross-correlation (XC) trace to a bi-exponential decay. The slow and fast components can be associated with e–e and e–ph interactions, i.e. internal thermalization and cooling of the electron gas, respectively*

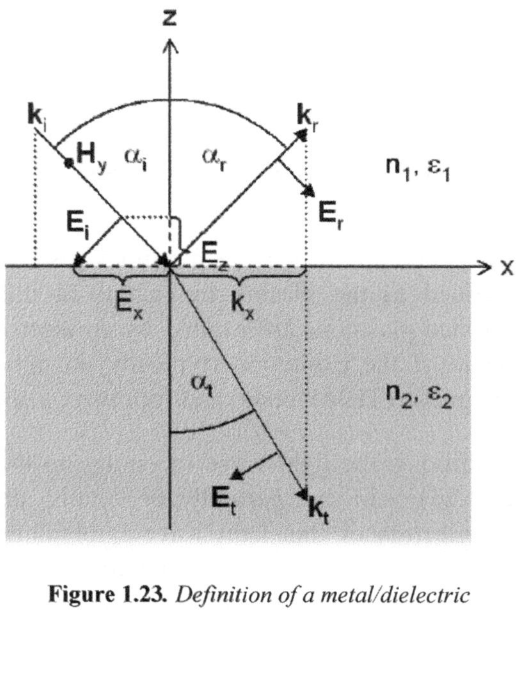

Figure 1.23. *Definition of a metal/dielectric*

The angle θ_c [1.53] corresponds to the total reflection angle defined by the index ratios n_m and n_d of the metal and dielectric material, respectively, forming the interface allowing the evanescent mode at the level of the interface and parallel to the x-axis

$$\theta_c = \sin^{-1}\left(\frac{n_m}{n_d}\right) \qquad [1.53]$$

Taking into account the permittivity of the dielectric ε_d, the oscillation frequency of the SPP ω_{sp} intervenes at a frequency lower than the plasma frequency ω_p [1.54]:

$$\omega_{sp} = \frac{\omega_p}{\sqrt{1+\varepsilon_d}} \qquad \omega_p = \sqrt{\frac{ne^2}{m\varepsilon_0}} \qquad [1.54]$$

It has been demonstrated that the component of the wave vector along the x-axis is defined by the following equation, representing the plasmonic dispersion equation [1.55]:

$$\boxed{k_x = \frac{\omega_0}{c}\sqrt{\frac{\varepsilon_d \varepsilon_m}{\varepsilon_d + \varepsilon_m}}} \qquad [1.55]$$

This dispersion equation, shown in Figures 1.24 and 1.25, allows us to understand that an optical coupling is needed in order to match the wave vector of the incident photon ($k_{photon,air}$ curve) with a plasmon present at the interface level of two dielectric materials with different optical index values (red and green curves).

A coupling via a prism or a grating (see Figure 1.26) adding a phase element in sin θ_r (θ_r representing the incidence angle in the prism or at the coupling grating level) is realized by natural reflection in the case of the prism and by disordered diffraction in the case of a periodic grating with a step Λ.

By considering the complex permittivity [1.56] of the metal:

$$\varepsilon_m = \varepsilon_m' + j\varepsilon_m'' \qquad [1.56]$$

Three conditions must be respected between this permittivity and the dielectric one:

$$\varepsilon_m' < 0$$

$$|\varepsilon_m'| \geq \varepsilon_d$$

$$\varepsilon_m'' \to 0$$

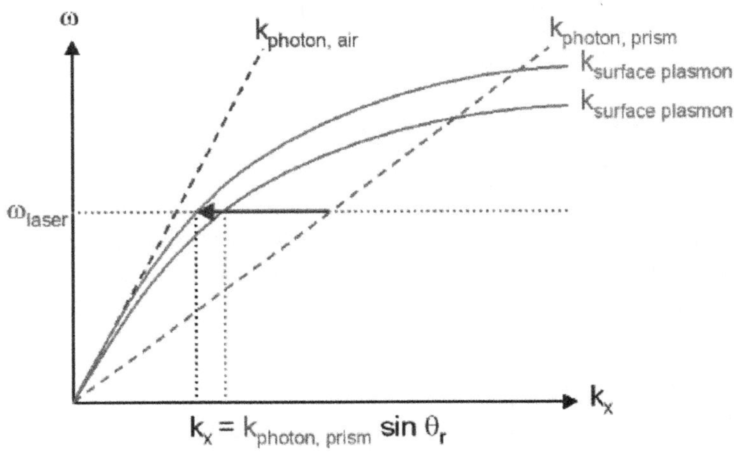

Figure 1.24. *Plasmonic dispersion relation for a prism coupling*

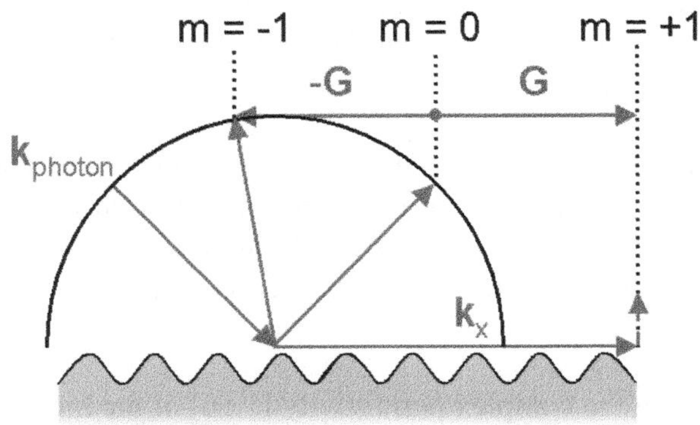

Figure 1.25. *Representation of a diffraction grating*

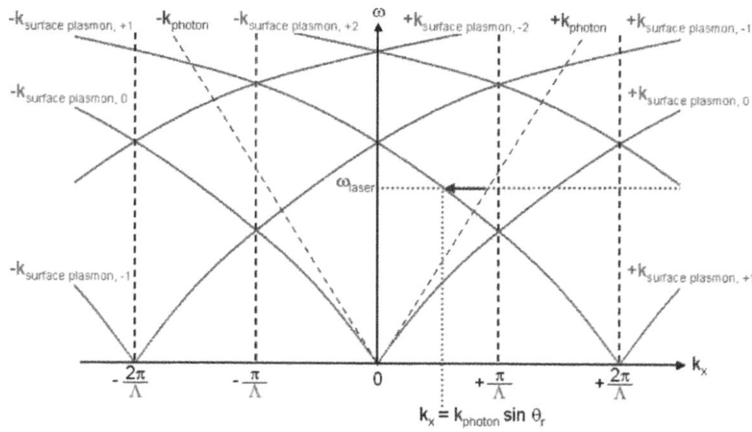

Figure 1.26. *Plasmonic dispersion relation in case of a coupling by grating*

1.3.2.2. *Matching conditions: metal/semiconductor interface*

In the conventional case, the generation of plasmon is done simultaneously between the two faces of the metal, i.e. at the coupling material/metal interface and the metal/dielectric interface inducing two potential configurations for the optical coupling either by the front face or by the rear face [KRE 71, OTT 68].

If the dielectric material is replaced by a semiconductor material for integration or efficiency reasons, a plasmon generation at the metal–semiconductor interface must be favored.

To satisfy this latest condition, constraints on materials must be taken into account. For example, it will be necessary to have an index of the coupling material greater than or equal to the semiconductor material index toward the desired incidence angle:

$$\varepsilon_{prism} >= \varepsilon_{dielectric}$$

2

Electromagnetic Material Characterization at Nanoscale

This chapter deals with the definition and parameters of some dedicated measurement techniques in order to define and measure the main parameters associated with a material. It addresses static and dynamic characterization techniques including ultrafast approaches both from the photonic and electronic points of view [SCH 06].

2.1. State of the art of macroscopic material characterization techniques in the microwave domain with dedicated equipment

2.1.1. *Static resistivity*

Determination of static resistivity ρ of a semiconducting material is one of the major key material parameters that allow its exploitation in the technology of any integrated device/circuit/system. Its knowledge will help to quantify the carrier density *n* and *p* and their associated mobility $\mu_{n,p}$ at thermodynamic equilibrium, as ρ is equal to $1/[q(n\mu_n + p\mu_p)]$. In this area, several characterization techniques today constitute state-of-the-art methods in a contacting or contactless electrical access environment, dedicated to 3D materials as described in the following.

2.1.1.1. *Two-point versus four-point probe*

The four-probe technique, originally proposed by Winner in 1916 in geophysics and then by Valdes in the case of semiconductor materials, is commonly used to determine the resistivity of materials using two identical probe terminals working in differential mode. In this configuration, parasitic voltage drops, as in contact resistances of metallic access pads, probe resistance and the spreading resistance made by current spreading at the end of probe, are canceled as each contact serves as a current and as a voltage probe leading to an absolute measurement of material resistance R_m by the knowledge of voltage V and current I (Figure 2.1).

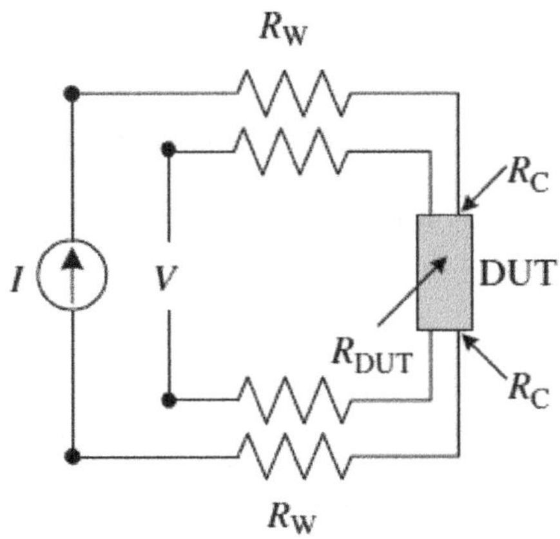

Figure 2.1. *Equivalent circuit of the four-probe measurement technique*

The 3D material resistivity value expressed in Ω·cm is statistically extracted from repetitive resistance measurements executed between different probe distances L, as R_m is equal to $\rho.\frac{L}{S}$, where S is referring to the 2D cross-section of a material piece existing between metallic pads. For most four-probe characterization, the probe distance is identical, leading to a direct calculation of resistivity as $\rho = 2\pi L \cdot V/I$.

Considering bulk to thin film 3D materials, correction factors must be applied for measurement accuracy.

2.1.1.2. *Resistivity profiling*

2.1.1.2.1. Differential Hall effect

The differential Hall effect is defined by the sheet resistance of a layer of thickness *t* over a substrate according to equation [2.1]

$$R_{sh} = \frac{1}{q\int_x [n(x)\mu_n(x) + p(x)\mu_p(x)]dx} \qquad [2.1]$$

If the sample is a thin layer, it must be separated from the substrate by an insulating layer to confine the four-point probe current to the layer.

2.1.1.2.2. Spreading resistance profiling

The spreading resistance probe technique is used today to generate resistivity and dopant density depth profiles. It has very high dynamic range (10^{12}–10^{21} cm^{-3}) and is capable of profiling very shallow junctions into the nanometer regime. Substantial progress has been made in data collection and treatment. The latter relates to improved sample preparation and probe conditioning procedures. The instrument consists of two carefully aligned probes that are stepped along the semiconductor surface. The resistance between the probes is given by $R = 2R_p + 2R_c + 2R_{sp}$, where R_p is the probe resistance, R_c is the contact resistance and R_{sp} is the spreading resistance. The resistance is measured at each location.

2.1.1.3. *Some contactless method examples*

Well adapted to nanotechnology, contactless resistivity measurement methods are divided into two broad categories: electrical and non-electrical measurements. Commercial equipment is available for both. Electrical contactless measurement techniques fall into several categories: the sample is placed into a microwave circuit and perturbs the transmission or reflection characteristics of a waveguide or cavity, the sample is capacitively coupled to the measuring apparatus, and the sample is inductively coupled to the apparatus.

Eddy current method is used to determine the resistivity using the integration of the material inside a resonant cavity. Induced Eddy currents by magnetic field excitation can be measured through Joule heating by energy absorption (equation [2.2])

$$P_a = K(V_T / n)^2 \int_0^t \sigma(x)dx \qquad [2.2]$$

2.1.2. *Carrier and doping density*

2.1.2.1. *Capacitance–voltage (C-V) and current–voltage (I-V) measurements*

Capacitance–voltage measurements are associated with semiconductor junctions where a space charge area is created and modulated by an applied voltage.

The C-V profiling method has been used with Schottky barrier diodes using metal, p-n junctions, metal-oxide-semiconductor (MOS) capacitors, metal-oxide-semiconductor field effect transistors (MOSFETs) and metal-air-semiconductor structures.

As far as nanotechnology is concerned, DC permittivity measurements have been performed through C-V characterization techniques applied to capacitive circuits. Specific attention must be given to the Debye length, which sets a limit to the spatial resolution of the measured profile. Indeed, this Debye length problem arises because the capacitance is determined by the movement of majority carriers, and the majority carrier distribution cannot follow abrupt spatial changes in doping density profiles.

2.1.2.2. *Hall effect*

The key feature of Hall measurements is the ability to determine the carrier density, carrier type and mobility.

Hall theory predicts the Hall coefficient R_H [2.3]

$$R_H = r(p - b^2 n)/[q(p + bn)^2] \qquad [2.3]$$

where $b = \mu_n/\mu_p$ and r is the scattering factor whose value lies between 1 and 2, depending on the scattering mechanism in the semiconductor.

The Hall coefficient is determined experimentally as in [2.4]

$$R_H = tV_H/BI \qquad [2.4]$$

where t is the sample thickness, V_H is the Hall voltage, B is the magnetic field and I is the current.

2.1.2.3. *Optical techniques*

Optical techniques become useful for material optical properties such as absorption coefficient determination and complex optical index over a wide high-frequency bandwidth. In this way, experimental techniques allow us to know the reflection, transmission and absorption coefficients R, T and A, respectively, by use of dedicated test cells in incidence with an optical signal.

The optical transmittance T [2.5] through a semiconductor wafer of thickness t is approximately

$$T = (1 . R)^2 .\exp(-\alpha t) \qquad [2.5]$$

For reasonable measurement sensitivity, αt should be of the order of unity. For α varying from 1 to 10 cm^{-1}, representative of shallow impurity absorption at low densities, the sample must be 1–10 mm thick. Samples of this thickness are convenient for bulk wafers but not for epitaxial layers, making the IR spectroscopy of thin layers impractical.

Other approaches such as photoluminescence are extremely helpful for the determination of the electronic band structure.

2.1.3. *Contact resistance and Schottky barriers*

According to the fact that all semiconductor devices have contacts and all contacts have contact resistance, it is important to characterize such contacts, especially in the case of 2D or 1D materials. Contacts are generally metal–semiconductor contacts, but they may be semiconductor–semiconductor contacts, where both semiconductors can be single crystal, polycrystalline or amorphous, linked to the identification of the Fermi level localization. For a discussion of the measurement techniques, the type of contact is unimportant, but the resistance of the contact material is important.

Several types of contact are conventionally studied: metal–semiconductor (Schottky) or ohmic.

The work function of a solid is defined as the energy difference between the vacuum level and the Fermi level. Work functions for the metal and the semiconductor are shown in Figure 2.2, with the metal work function φ_M being less than the semiconductor work function φ_S. The work function is given as the energy φ_M related to the potential φ_M by $\varphi_M = \varphi_M/q$.

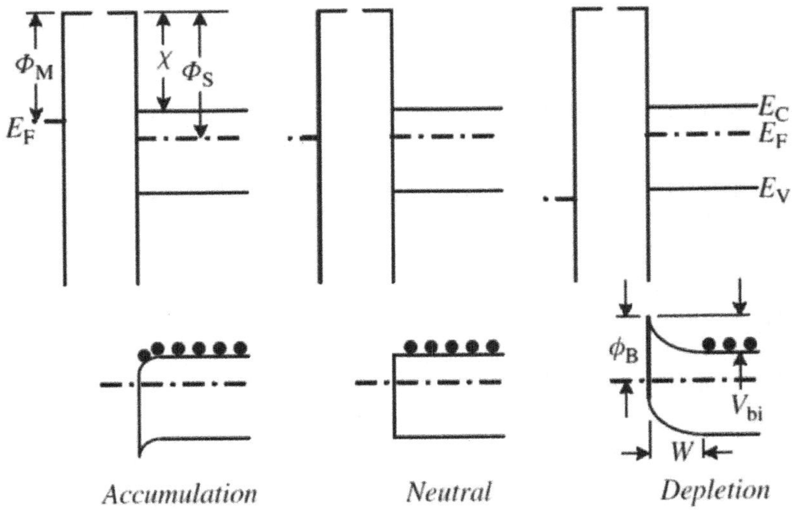

Figure 2.2. *Schematic of the metal–semiconductor contacts according to the Schottky model*

2.1.4. *Transient methods for the determination of carrier dynamics*

2.1.4.1. *Recombination lifetime – optical measurements*

Lifetimes are divided into two primary categories: recombination lifetimes and generation lifetimes. The concept of the recombination lifetime τ_r holds when excess carriers decay as a result of recombination. When these recombination and generation events occur in the bulk, they are characterized by τ_r and τ_g. When they occur at the surface, they are characterized by the surface recombination velocity v_r and the surface generation velocity v_g. Both bulk and surface recombination or generation occur simultaneously and their separation is sometimes quite difficult. The measured lifetimes are always effective lifetimes consisting of bulk and surface components.

Pump–probe measurement techniques, using a combination of electro-optic probes detecting ultrashort electrical signals, have been widely developed for identification of ultrafast carrier lifetimes in 3D, 2D and 1D materials such as in [ORT 01, COU 08].

2.1.4.2. *Conductivity mobility*

The study and applications of the Hall effect can be found in the literature focused on solid state physics and semiconducting materials. The Hall effect measurement technique has found wide application in the characterization of semiconductor materials because it gives the resistivity, carrier density and mobility. Hall found that a magnetic field applied to a conductor perpendicular to the direction of current flow produces an electric field perpendicular to the magnetic field and the current. For more detail on this, see section 1.3.1.

2.1.5. *Frequency methods for complex permittivity determination in frequency*

2.1.5.1. *Scientific background*

The principles and techniques of permittivity measurements using transmission lines have been widely studied for the determination of bulk material permittivity, permeability and conductivity in the frequency domain. A non-destructive method by electromagnetic coupling of an unknown material thin layer with a tranverse electromagnetic (quasi-TEM) propagation mode in planar configuration (microstrip or coplanar line) has become the most appropriate methodology today to achieve a complete and broadband characterization [CHE 10, KAS 10]. For nanomaterial characterization, only coaxial line techniques have been used because of CNT delivery in disperse solutions [DEC 12] inducing frequency bandwidth limitation and the random orientation of CNTs compared to the E-field polarization direction. In this chapter, we present a new measurement technique of frequency-dependent complex anisotropic permittivity of in-situ grown nanomaterials by integration in a planar multilayered structure.

2.1.5.2. *Direct extraction of the propagation constant of a coplanar multilayered structure*

After an analytic S-to-ABCD-matrix conversion, equations [2.6] and [2.7], and a de-embedding procedure for microwave probe parasitic perturbation suppression, exact determination of the effective propagation constant (γ_{reff}) [2.8] and permittivity (ε_{reff}) [2.9] of a transverse N-layer structure of length L is obtained.

$$\begin{pmatrix} A & B \\ C & D \end{pmatrix} = \begin{pmatrix} \frac{(1+s_{11})(1-s_{22})+s_{12}s_{21}}{2s_{21}} & Z_0 \frac{(1+s_{11})(1+s_{22})-s_{12}s_{21}}{2s_{21}} \\ \frac{1}{Z_0} \frac{(1-s_{11})(1-s_{22})-s_{12}s_{21}}{2s_{21}} & \frac{(1-s_{11})(1+s_{22})+s_{12}s_{21}}{2s_{21}} \end{pmatrix} \quad [2.6]$$

$$\text{with} \begin{pmatrix} A & B \\ C & D \end{pmatrix} = \begin{pmatrix} cosh(\gamma_{eff}.L) & jZ_c.sinh(\gamma_{eff}.L) \\ \frac{j}{Z_c}.sinh(\gamma_{eff}.L) & cosh(\gamma_{eff}.L) \end{pmatrix} \quad [2.7]$$

$$\gamma_{reff} = \frac{cosh^{-1}(A)}{L} \quad [2.8]$$

$$\varepsilon_{reff} = \frac{\gamma_{reff}^{\ 2}}{j^2\omega^2c^2\mu_{reff}} \quad [2.9]$$

2.1.5.3. *Permittivity extraction of unknown layer by inverse problem resolution*

As an inverse problem, relative permittivity of the SWCNT layer extraction is achieved using a conformal mapping technique from the Veyres–Fouad Hanna approximation [VEY 80], assuming a multilayered structure as a superposition of capacitances C_i of each individual material layer i with coplanar transmission line (Figure 2.3). The total capacitance C is equal to the sum of each individual capacitance [2.10].

$$C = \sum_{i=0}^{N} C_i \quad [2.10]$$

Individual capacitance C_i of layer i is directly related to the complete integral equations $K(k_i)$ and $K'(k_i)$ expressed in equations [2.11] and [2.12] assuming the identity $K(k'_i) = K'(k_i)$.

$$C_i = 2\varepsilon_0\left(\varepsilon_{ri}^{S,T} - \varepsilon_{r_{i-1}}^{S,T}\right)\frac{K'(k_i)}{K(k_i)} \quad [2.11]$$

$$\text{with } k_i = \frac{sinh\left(\frac{\pi x_c}{2h_i}\right)}{sinh\left(\frac{\pi x_b}{2h_i}\right)} \cdot \sqrt{\frac{sinh^2\left(\frac{\pi x_b}{2h_i}\right) - sinh^2\left(\frac{\pi x_a}{2h_i}\right)}{sinh^2\left(\frac{\pi x_c}{2h_i}\right) - sinh^2\left(\frac{\pi x_a}{2h_i}\right)}}$$

$$\text{and } k'_i = \sqrt{1 - k_i^2} \quad [2.12]$$

Specific expressions must be applied for layer 0 representing the coplanar line capacitance in air as expressed in [2.13].

$$C_0 = 4\varepsilon_0 \frac{K\prime(k)}{K(k)} \text{ with } k_0 = \frac{x_c}{x_b} \cdot \sqrt{\frac{x_b^2 - x_a^2}{x_c^2 - x_a^2}} \text{ and } k'_0 = \sqrt{1 - k^2{}_0} \qquad [2.13]$$

From the determination of effective permittivity $\varepsilon_{\text{reff}}$ of the complete multilayered structure, layer characteristics such as thickness, coplanar line dimensions and relative permittivity of one layer can be extracted by analytic resolution of equation [2.10] in its completed expression.

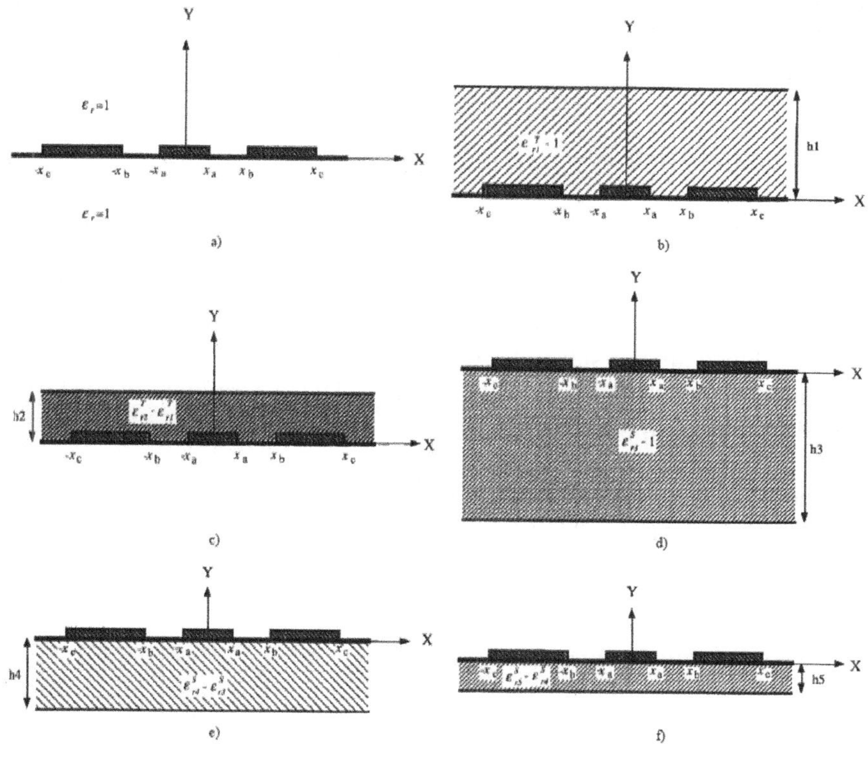

Figure 2.3. *Transverse representation of a five-layer structure for individual capacitance C_i definition from [PIE 10a]: (a) C_0, (b) C_1, (c) C_2, (d) C_3, (e) C_4 and (f) C_5*

2.2. Evolution of techniques for nanomaterial characterization

For 1D materials, the problems associated with attempting to scale down traditional semiconductor devices have led researchers to look into CNT-based devices, such as CNT FETs, as alternatives. Because they are not subject to the same scaling problems as traditional semiconductor devices, CNT FETs are being studied for a wide variety of applications, including logic devices, memory devices and sensors.

The research on these devices typically involves determining various electrical parameters, which may include current–voltage (I-V), pulsed I-V and capacitance (C) measurements. Characterizing the electrical properties of delicate nanoelectronic devices requires instruments and measurement techniques optimized for low power levels and high measurement sensitivity.

2.2.1. *The CNT transistor*

A single semiconducting CNT can be used as the conducting channel between the source and the drain of an FET. Two metal contacts are located across both ends of the CNT to form the source and drain terminals of the FET. The CNT is placed at the top of an oxide layer that has been processed above a doped silicon substrate, which forms the gate terminal. Connections are made to the three device under test (DUT) terminals to perform the electrical measurements.

The I-V characteristics of a CNT transistor can be used to extract many of the device's parameters, study the effects of fabrication technique and process variations, determine the quality of the contacts, etc.

2.2.2. *Optimizing DC measurements*

The following techniques will improve the quality of DC measurements made on CNT FETs:

1) Limit current: to prevent damage to the device while performing I-V characterization, the user should limit the amount of current that can flow through the device.

2) Provide sufficient settling time: because CNT FET measurements often involve measuring low current (<1 μA), it is important to allow

sufficient settling time to ensure the measurements are stabilized after a current or voltage has been applied. Some of the factors that affect the settling time of the measurement circuit include the cables, test fixtures, switches, probers, the DUT resistance and the current range of the measurement instrument. To ensure settled readings, additional delay time can be added to the voltage or current step time before the measurement.

3) Use proper speed modes: this affects the settling time of the reading, as well as the integration time of the measurement.

4) Minimize noisy measurements: noise may be generated from a variety of sources, including particle collisions, defects, AC pick-up and electrostatic interference. Noisy measurements result when a noise signal is superimposed on the DC signal being measured. This can result in inaccurate or fluctuating measurements. Electrostatic interference is another cause of noisy measurements when measuring low currents. This coupling occurs when an electrically charged object approaches the circuit under test. In high-impedance circuits, this charge does not decay rapidly and can result in unstable measurements.

2.2.3. *Pulsed I-V measurements*

In addition to making traditional DC I-V measurements, it may be desirable to perform ultrafast pulsed I-V measurements for various reasons. First, it may be important to observe the high-speed response of the CNT device. In some cases, nanostructures can be destroyed by the heat generated when making traditional DC measurements. Pulsed I-V measurements can reduce the total energy dissipated in a device and therefore reduce the potential for damage. Finally, pulsed electrical testing can prevent current drifting in measurements that can occur during DC measurements.

2.2.4. *Capacitance–voltage measurements*

In addition to performing DC and pulsed I-V measurements on CNT FETs, measuring the capacitance of the FET can also provide information about the device, including the mobility, timing effects and gate dielectrics.

2.2.4.1. *Optimizing capacitance measurements*

To improve the quality of capacitance measurements, it will be necessary to perform open compensation (for measurements < 10 pF).

2.3. Micro- to nanoexperimental techniques for the characterization of 2D, 1D and 0D materials

In the field of electrostatic force microscopy (EFM) of nanostructures, experiments aiming to probe the local electronic properties of nanostructures on insulators have been realized. Such experiments are based on the use of an AFM operated in one of its electrical modes: EFM, Kelvin force microscopy (KFM), etc. Results based on charge injection and EFM experiments applied to 0D and 1D nanostructures have been obtained on silicon nanoparticles and CNTs. Experimentally, the tip of the EFM cantilever is used to address a single nanostructure and inject charges when pressed with a few nanonewtons of force and biased with respect to the substrate at an injection voltage (Figure 2.4). In this process, charges are transferred (in most cases from the EFM tip into the nanostructure). EFM data consist of the shift of the resonance frequency f0 of the cantilever, when biased at a detection voltage VE_{FM}. EFM frequency shifts are proportional to the gradient of the electrostatic force acting on the EFM tip [MÉL 04].

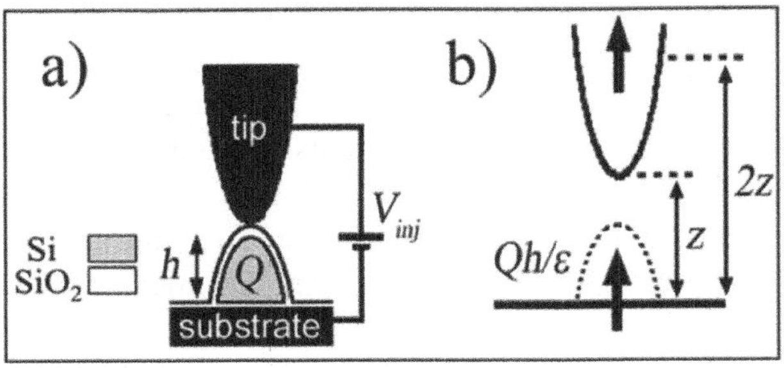

Figure 2.4. *a) Injection of an amount Q of charges in a single silicon nanoparticle (gray) of height h by tunneling through the nanoparticle native oxide (white) from the EFM tip biased at Vinj. b) Equivalent surface dipole Q x h/ε. When the tip is scanned at a distance z over the substrate plane, the dipole interacts with its tip image. In a plane geometry, this image dipole of momentum Q x h/ε would be located at 2z over the substrate plane*

Specific techniques have recently been developed in order to keep micrometer scale access for nanometer scale material parameter extraction, from DC to microwave frequencies. As high impedance is mainly demonstrated by such materials in contact with electrical electrodes, an electrical access environment has to be reconsidered in order to overcome high contact resistance. In such a way, electrical circuits relying on a Wheatstone bridge implementation could be a technical solution [CUR 08].

3

Nanotechnology-based Components and Devices

This chapter summarizes some research approaches for the realization of new optically controlled microwave functions. This concept, part of the microwave photonics research area, driving a new family of microwave devices, mainly deals with the photoconductivity in semiconductors with 1D, 2D and 3D structures for applications ranging from the generation and sampling of signals from the microwaves to the THz frequencies.

This chapter presents the newly envisaged optically controlled microwave functions at nanoscale and describes the main parameters involved in these nanodevices and associated functions.

A photoconductive switch is an electrical switch, which is based on the photoconductivity of a material, i.e. an increase in its electrical conductance as a result of illumination by light. In almost all research in this field, a semiconductor material is used, where the absorbed light (with a photon energy above the bandgap energy) generates electrons and holes carriers, which then contribute to the conductivity.

In the last decades few decades, the most commonly used semiconductors were chromium-doped gallium arsenide (Cr-GaAs), low-temperature-grown gallium arsenide (LT-GaAs), indium phosphide (InP), cadmium selenide (CdSe), and Si and its derivatives (amorphous silicon and silicon on sapphire (SoS)).

To reduce the recovery time of the switch (determined by the lifetime of photoexcited carriers), we typically use low-temperature growth (often followed by rapid thermal annealing), some doping (e.g. chromium in GaAs) or ion bombardment for producing crystal defects. Apart from the recovery time, important criteria are the bandgap energy, dark resistivity and electrical breakdown resistance.

There are different designs of photoconductive switches:

– bulk devices, which are several millimeters or even centimeters long with electrical contacts on the end faces, used for switching very high voltages (sometimes above 100 kV);

– devices with a small gap in a microstrip; the gap can be straight or interdigitated and has a width between a few microns and tens of microns for low-power applications with very high speed;

– *sliding contact devices* for the highest speed, where a point between the two parallel strips of a coplanar stripline is illuminated.

All such devices are of the metal–semiconductor–metal (MSM) type.

The main applications of photoconductive switches are as follows:

– sampling of analog signals in very fast analog-to-digital converters;

– generation of submillimeter wave and terahertz pulses;

– generation of microwaves and millimeter waves.

The new generation of ultra-high-speed photoconductive switching started on the basis of the Auston photo-switch [AVS 83] using LT-GaAs due to its ultrafast trapping characteristics and low leakage current [MCL 97, TAN 96]. However, at the end of 20th century only high-quality photoconductive materials requiring an 800 nm illumination exist to switch such devices resulting in the use of bulky and expensive Ti-Sapphire pulse lasers. On the contrary, 1550 nm wavelength optical control through the availability of compact fiber and solid-state mode locked lasers sources was accessible but without high-quality photoconductive materials.

3.1. Photoconductive switches for microwave applications

3.1.1. *Major stakes*

Integration of optical technologies into telecommunication systems enabled the improvement of performances together with a reduction of the dimensions and architecture complexity of these systems. One aspect of integration is addressed in the direct optical control of microwave circuits in integrated technology. The optical signal is used to command either the active circuits or passive devices.

Concerning the control of nonlinear circuits, the optical signal is applied directly to their active components. As an example, and following the approaches, the optical control of the gain of a microwave amplifier and the optical control of a microwave oscillator frequency and output power were validated [DE 81, YEN 77, FAC 05]. The main drawback of these circuits is linked to the low dynamic range of the optical control. For the control of linear microwave circuits, with transmission line discontinuity, functions such as switching and phase shifting based on photoconductivity have been demonstrated [TRI 12, CAN 03]. Photo switching gives rise to research aiming at THz signals generation and sampling. Optically controlled phase shifting opens new horizons in the achievement of new wideband phase array antenna architecture in the transmit and receive modes. Moreover, optically tunable filter and resonator functions have been realized.

3.1.2. *Basic principles*

The simplest optically controlled microwave device is called the photoconductive switch or photoswitch. It is based on a metallic microwave transmission line interrupted by a gap of length L_g and width w, deposited on a semiconductor substrate of thickness h and relative permittivity ε_r.

The gap between the two electrodes will act as a light/semiconductor interface, i.e. it will allow the illumination, by an optical source, of the semiconductor substrate that gives rise to the photoconductor effect (Figure 3.1).

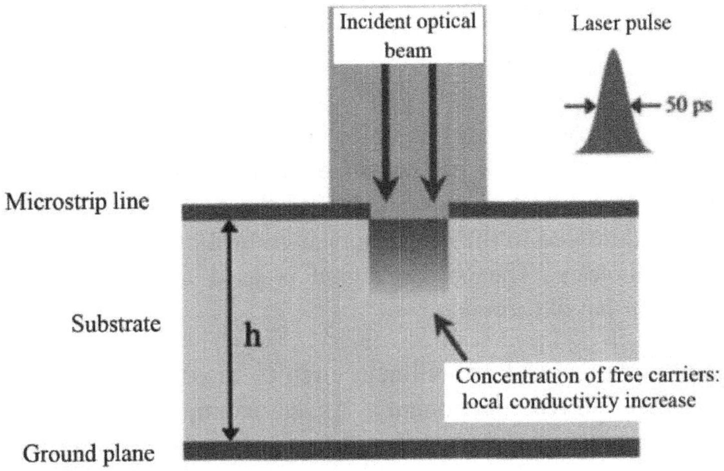

Figure 3.1. *Schematic of a cross-section of a microwave photoswitch in a microstrip configuration*

This approach is a basic building block necessary to realize optically controlled microwave monolithic integrated circuits (O-MMICs) for wideband applications. It is designed for any semiconductor material, depending on the applications in terms of operating frequencies and achievable optical wavelength.

In the field of microwaves, two main MMIC technologies have been developed: coplanar waveguide (CPW) and microstrip configurations (see Figure 3.2).

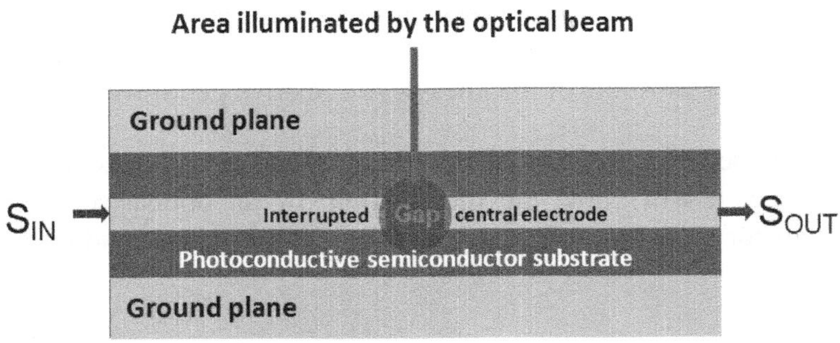

Figure 3.2. *Schematic of the interrupted coplanar waveguide configuration*

The operating principle is as follows:

– Without an optical illumination (OFF state), the propagation of the microwave signal is almost completely interrupted by the slit of the line. The associated device is electrically equivalent to a capacity in series C_g associated with the electrodes created by the transmission line interruption (slit), and also with two capacities in parallel, associated with the coupling of the electrodes to the ground plane C_p.

– Under light illumination (ON state), the photon energy is absorbed by the semiconductor material inducing a reduction of the resistivity due to the creation of a localized free-carrier plasma area, characterized by the photoconductance parameter extracted from equation [3.1]. The propagation of the microwave signal through the line slit is thus allowed. As a result, a photoresistance is added in parallel to the capacitance C_g in order to complete the equivalent electrical scheme (see Figure 3.3).

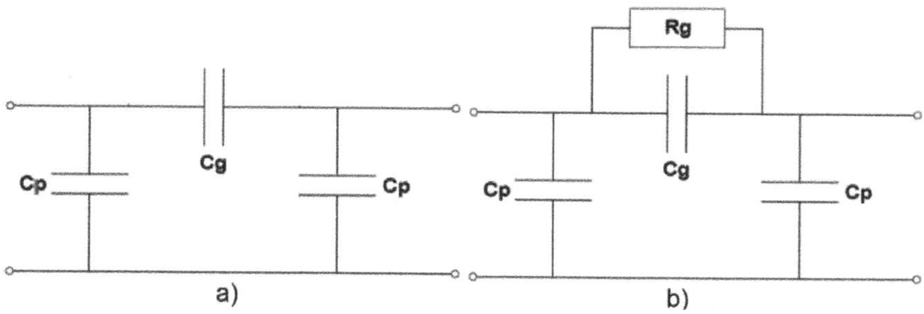

Figure 3.3. *Electrical equivalent scheme of a microwave photoswitch a) in the OFF state and b) in the ON state*

The associated device works as a switch, modifying the phase and amplitude of the microwave signal flowing through this electrical discontinuity.

The ability of this microwave photoswitch to enable or to prevent the transmission of a microwave signal is defined by the complex variable $\Re_{ON/OFF}$ representing the ON/OFF ratio, determined from the transmission S parameters, S_{21} (or S_{12}), in these two states, as defined in equations [3.2] and [3.3]. The insertion losses of the device are determined by the module of the S_{21} (or S_{12}) parameters, obtained through light illumination and expressed in decibels.

In these equations, the parameters are the following:

– μ_n and μ_p represent, respectively, the carrier mobilities of the electrons and holes photogenerated;

– η is the quantum efficiency;

– α is the absorption coefficient;

– R is the reflection coefficient at the air/semiconductor interface;

– P_{opt} is the incident optical power;

A and w_{eff} are the illuminated area and the diameter of the optical beam;

– L_g is the effective length of the illuminated area (which could differ from w_{eff});

– L is the ambipolar diffusion length of the carriers;

– v_s is the recombination speed at the surface level and τ is the carrier lifetime.

$$G_g = \frac{e}{h\nu}(\mu_n + \mu_p)\,\eta\,\alpha\,\tau(1-R) \times \frac{P_{opt}}{A}\frac{W_{eff}}{L_g}\frac{1}{(1-\alpha^2L^2)}\left[\frac{1}{\alpha} - L\frac{\alpha L^2 + v_s\tau}{L + v_s\tau}\right] \quad [3.1]$$

$$\Re_{On/Off} = S_{21}(On)/S_{21}(Off) = R_{On/Off}\, e^{i\Delta\varphi_{On/Off}} \quad [3.2]$$

$$S_{21}(Off) = j2\omega Z_0 Cg/(1 + j2\omega Z_0 Cg)$$

$$S_{21}(On) = \frac{2Z_0 Gg(1 + j\omega\, Cg/Gg)}{(1 + 2Z_0 Gg)(1 + j\,\omega 2 Z_0\, Cg/(1 + 2Z_0\, Gg))} \quad [3.3]$$

3.1.3. *State of the art of photoconductive switching*

For decades, numerous research activities have been driven by improving the performances of photoconductive switching devices due to the results obtained with the new growth technologies of semiconductors. These new approaches offer the ability to realize components and devices with absorption capabilities in spectral regions compatible with the optical sources, i.e. in the visible domain (from 532 to 670 nm) and in the near IR (0.78–1.55 μm). Nowadays, the most studied materials are Gallium Arsenide (GaAs) and Indium Phosphide (InP) in the III-V for ultra-high speed applications and Silicon (Si)-based for the applications in the millisecond (ms) range semiconductor compounds families.

3.1.3.1. *Switching of continuous wave (CW) electrical signals by a CW optical source*

As mentioned earlier, the first photoswitch was demonstrated by D.H. Auston in 1975 by realizing the switching of a continuous signal of 20 V by an optical beam in free-space propagation. The illuminating optical signal was obtained at 0.53 μm after a frequency doubling of 1.06 μm laser source based on a potassium di-hydrogen phosphate (KDP) crystal. This laser source at 0.53 μm illuminates a Si-based microstrip realizing the ON state. For the OFF state, the 1.06 μm source was used to realize a direct contact to the ground plane.

3.1.3.2. *Switching of CW electrical signals by an intensity-modulated optical source*

An interrupted CPW gap illuminated by an impulse optical source was demonstrated [CAS 76]. The photoconduction is obtained by applying an optical pulsed signal between the transmission line central electrode and the ground plane.

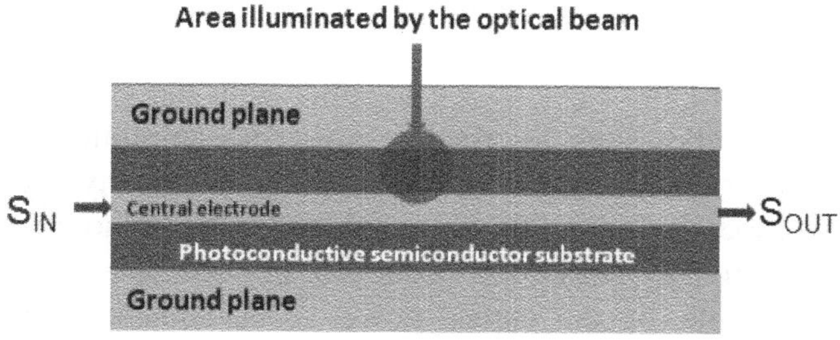

Figure 3.4. *Schematic of the short-circuited coplanar waveguide configuration*

By integrating a thin plate of GaAs:Cr into a coaxial line charged to 50 Ω, it was possible to switch a CW signal of 250 V with a pulse of 30 ps and 10 μJ.

This approach opens the way to switch high-voltage signals by ultrashort optical pulses in order to realize high-power microwave signals.

This approach requires a semiconductor material with an ultrashort carrier recombination time. This leads to the studies of doped materials or low-temperature grown materials such as GaAs with carrier lifetime in the order of a few picoseconds [LID 92] or below one picosecond [LIN 12], measured by temporal autocorrelation. Another attempt of the same performances was achieved by using GaAs: As+ [LIN 97].

3.1.3.3. *Switching microwave signals by a CW optical source*

Numerous semiconductor materials were studied for switching microwave signals such as amorphous Si, Ge, C diamond, CdS, CdS0.5Se0.5, GaP and SoS.

An interdigitated gap photoconductive microwave switch using InP:Fe under an optical illumination at 0.805 μm was realized, showing an ON/OFF ratio of 43 dB at 100 MHz and 23 dB at 1 GHz under 50 mW optical power [AND 89].

3.1.4. *Photoconductive switching at nanoscale – examples*

Up to now, the best performances in terms of the ON/OFF ratio of microwave frequencies have been obtained due to the use of high-resistivity materials ($\rho > 10^7$ Ω·cm) and the extreme confinement of the area of interaction between the optical illumination and the semiconductor. This approach requires an optimization of the dimension in the submicron region.

The reduction of the capacity Cg together with the confinement of the illuminated area allows us to greatly lower the amplitude of the radio frequency (RF) signal in the OFF state while drastically increasing the carrier density locally in the ON state, inducing a large modification to the material conductivity. These last achieved performances have a strong impact, for example, on the optical power to be launched for the optical control of the photoswitch.

Today, thanks to the development of technological tools dedicated to nanotechnology, the definition of nano scale interaction volume by the reduction of microwave transmission lines with optical waveguides to the nano scale (Figure 3.5 – Table 3.1) is reachable.

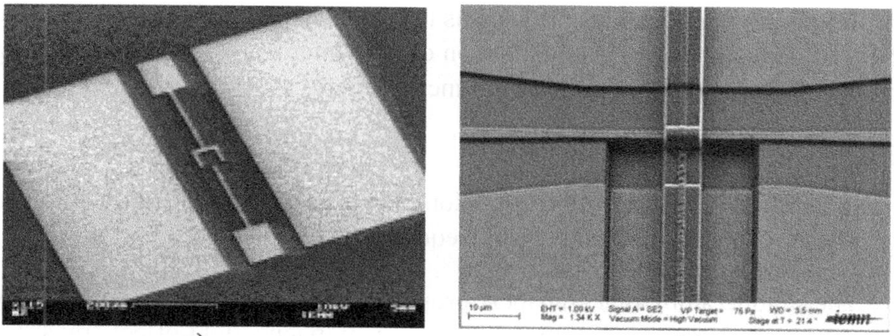

a) b)

Figure 3.5. *Examples of CPW-based microwave photoswitch on LT-GaAs; a) with an external illumination [TRI 06] or b) with an integrated optical waveguide (courtesy of IEMN)*

Key parameters	Dimensions	
	Micron scale	Submicron scale
Interaction volume	20 x 20 x 2 μm^3	1 x 1 x 0.5 μm^3
Insertion losses (@ 20 GHz)	30 dB	<10 dB
ON/OFF ratio (amplitude) (@ 20 GHz)	~ 3 dB	>30 dB

Table 3.1. *Performances of GaAs nanophotoswitch under optical illumination at 0.8 μm*

3.2. 2D materials for microwave applications

3.2.1. *Graphene for RF applications*

As mentioned in the Introduction, recent results on the use of graphene for microwave applications allowed us to revisit some functions covering the field of nanocircuits for the realization of new microwave functions. Among them, we will review the following functions here.

3.2.1.1. *RF mixers and multipliers*

Recently, IBM showed the operation of transistors fabricated on epitaxial graphene with a current gain cutoff frequency f_T of 100 GHz.

The unique ambipolar transport properties of graphene are explained by the fact that the electrical conduction in a graphene field-effect transistor (GFET) can be dominated by either positive holes or negative electrons, depending on the bias applied to the gate electrode. It has been shown that the resulting effect is a V-shaped drain current–gate voltage (ID–VGS) characteristics [WAN 10], combined with its extremely high mobility. This will allow the development of a new form of nonlinear electronics for RF and mixed-signal applications [MER 08].

As an example of these new nonlinear devices, frequency doubling has been realized with a single graphene transistor by biasing the gate to the minimum conduction point and superimposing a sinusoidal input signal to the gate. In 2009, demonstration of a new frequency multiplier device based on graphene with the potential has been made in order to overcome the main problems of conventional frequency multipliers. For the first time, frequency doubling is realized with just a single transistor device that can give very high spectral purity at the output without any additional filtering [WAN 09].

3.2.1.2. *Terahertz absorber*

In the infrared-to-visible spectral range, the expected absorbance has been calculated to be independent of frequency and to have a magnitude given by $\pi\alpha\sim$ 2:293% and validated experimentally [MAK 08]. This extremely large value of absorbance for monoatomic layer materials opens the way to new far-IR and THz devices such as perfect absorber in the THz region [AND 13].

3.2.1.3. *Isolator*

The anisotropic tensor conductivity of a magnetically biased graphene sheet can be used to introduce Faraday rotation. Thus, a magnetically biased graphene sheet can be used to rotate the polarization of a normally incident plane wave. As an application of the graphene Faraday rotation, they propose a circular waveguide Faraday rotator. It consists of a circular cylindrical waveguide loaded with a graphene sheet perpendicular to its axis. The main advantage of this configuration compared to conventional configurations based on ferrites is its bandwidth, which is limited only by the waveguide, since graphene is almost frequency independent at microwave frequencies [ATT 12].

Furthermore, the extremely thin nature of graphene (one atomic layer, 1.42 A°) provides exceptional compactness and better heat sinking compared to ferrite solutions. The power handling might also be excellent, due to the unique thermal properties of graphene up to 400 K. These non-reciprocal gyrotropic properties of graphene can also be used for the realization of a non-reciprocal spatial isolator [SOU 12].

3.2.1.4. *RF switches*

Graphene, due to its large Young modulus, appears as a potential material for nanoelectromechanical (NEMS) devices. Dragoman *et al.* [DRA 07] showed that graphene can also act as a very efficient RF-NEMS switch for microwave applications. In this case, metallic graphene nanomembranes (a single graphene monolayer with a thickness of l nm) are periodically loading a CPW. It is found that the resistance is tunable as a function of the applied DC voltage in the 30–100 V range while the capacitance was approximately 1 pF. This simple structure can be used as a tunable voltage variable attenuator in microwave range. The RF MEMS switch is a good candidate for applications working in a wide bandwidth frequency range up to 60 GHz.

3.2.1.5. *Transistors and field effect transistors (FETs)*

The first graphene transistor was reported by AMICA and Aachen University in 2007. In the following 3 years, the cutoff frequency of GFET transistors has increased from a few gigahertz to 100 GHz. The cutoff frequency of this device is as high as 170 GHz at a drain voltage of 2.2 V. These results unveil the great potential of graphene for future RF applications.

Numerical simulations predict that the cutoff frequencies could be in the order of several THz. Fabrication and experimental characterization of different GFETs have been introduced by different authors based on different fabrication technologies and facilities.

The graphene transistors also possess high current density of > 1 mA/μm.

One of the most important properties of graphene is a strong electric field effect, which leads to a carrier density in the range of $n < 10^{14}$ cm^{-2}, with high carrier mobilities for both electrons and holes (as high as 10^4 cm²/Vs at room temperature), this attracts a lot of attention to graphene as a possible material for a future high-speed field effect transistor (FET).

The most often studied GFET structure is a back-gated configuration where the graphene flake is contacted to form source and drain electrodes and the substrate acts as a back-gate. By depositing a dielectric layer on top of such devices, we can achieve a top-gate configuration allowing both gate biases to control the charge concentration in the device channel. Nevertheless, graphene was found to be attractive for high-speed analog electronics, where transistor current gain is more important than I_{ON}/I_{OFF}. Transistors with cutoff frequencies as high as 350 GHz and high-frequency circuits have recently been demonstrated.

3.2.1.6. *Graphene barristor*

Researchers from Samsung demonstrated the first graphene-based triode device called barrister. After a three-terminal active device, a graphene variable barrier "barristor" (GB), in which the key is an atomically sharp interface between graphene and hydrogenated silicon. Large modulation on the device current (ON/OFF ratio of 10^5) is achieved by adjusting the gate voltage to control the graphene-silicon Schottky barrier [YAN 02].

3.2.2. *Optoelectronic functions*

3.2.2.1. *Graphene photodetectors*

Technical challenges concern the achievement of new graphene-on-Silicon photodetector and its integration on chips to obtain graphene-based optoelectronic chips for more efficient broadband optical signal processing [KLE 13]. Through experiments, some researchers found that their detector would generate, without bias, 16 mA of current for each watt of incoming light. Its detection frequency was 20 GHz, already competitive with germanium.

Demonstration of an ultrawideband CMOS-compatible photodetector based on graphene. A multigigahertz operation over all fiber-optic telecommunication bands has been demonstrated. Further optimization of the device geometry led to high-speed electrical measurements on graphene-based photodetectors operating at frequencies up to 40 GHz [POS 13].

Demonstration of a graphene/silicon-heterostructure photodiode formed by integrating graphene onto a silicon optical waveguide on a silicon on insulator (SOI) with a near- to mid-IR operational range has been achieved [WAN 13]. The waveguide enables the absorption of evanescent light that propagates parallel to the graphene sheet, which results in a responsivity as high as 0.1 A/W at a 1.5 V bias at room temperature.

Photoconductivity in graphene devices has been studied widely across the spectral range from visible light to telecommunication bands.

As explained in Chapter 1, graphene demonstrates ultrafast carrier dynamics for both electrons and holes, and it has been shown that a weak internal electric field allows high-speed and efficient photocarrier separation and the weak optical absorption of graphene limits its photoresponsivity [GAN 13]. A waveguide-integrated graphene photodetector that simultaneously exhibits high responsivity, high speed and broad spectral bandwidth has been demonstrated. Using a metal-doped graphene junction coupled evanescently to the waveguide, the detector achieves a photoresponsivity exceeding 0.1 A/W together with a nearly uniform response between 1,450 and 1,590 nm.

3.2.3. *Other potential applications of graphene*

Another route to graphene-based electronics [OBE 11] is to consider graphene as a conductive sheet rather than a channel material that can be used to make:

– a single electron-transistor (SET);

– superconducting FETs and room temperature spintronics;

– transparent electrodes.

3.3. 1D materials for RF electronics and photonics

3.3.1. *Carbon nanotubes in microwave and RF circuits*

Ziaei [ZIA 08] introduced an RF nanoswitch based on vertically aligned carbon nanotubes (CNTs). It consists of CNTs perpendicular to the substrate. Two different architectures are proposed for this CNT switch: a series-based switch using ohmic contact between CNTs and a capacitive-based switch implemented in shunt configuration. The RF ohmic switch is designed by implementing CNTs in two sides of a CPW discontinuity. By applying DC voltage on the two sides of the CPW discontinuity, an electrostatic force is introduced between the two arms of the CNT switch. By electrostatic attraction, the switch is electrically closed and the RF signal is transmitted across the CPW. On the other hand, the shunt switch is based on two nanotube capacitive contacts between the inner line and the two sides of the ground planes of the CPW.

For radiation applications, since 2005 Rutherglen *et al.* [RUT 09] have presented a detailed study on CNT dipole antenna based on simple transmission line approximation.

Most recent nanotechnology developments in RF and THz electronics have been reviewed by Cha *et al.* [CHA 11]. On the basis of the equivalent circuit model of a single-walled CNT (SWCNT) transistor [BUR 04], it is predicted that a cutoff frequency of THz should be achievable.

Other interesting configurations for active components based on CNT are cold cathode THz vacuum tubes, which were proposed by the National Aeronautics and Space Administration (NASA) Jet Propulsion Laboratory (JPL).

CNT has a potential to be used as nanoscale transmission lines and as high-performance passive components in nanoelectronics [ZHA 06].

To distinguish the resistance from the imperfect contact or carrier scattering in the diffusive region, we present a resistance extraction methodology for a single CNT. The resistance of a single CNT consists of the following three parts:

– the intrinsic quantum resistance $R_0 = h/(2e^2)$ resulting from the carrier scattering between a 1D and 3D conductor;

– the resistance in the diffusion region caused by phonon scattering $hL/(4e^2\lambda_{eff})$, where L is the length and λ_{eff} is the effective mean free path (MFP) of the CNT.

– the contact resistance between the CNT and the metal $R_{contact}$.

The quantum resistance and the resistance caused by phonon scattering are related to the conduction channels $N(d)$, which depend on the diameter of the CNT.

The total resistance of a single CNT is described by $R_1 = 2R_{contact} + R_0(1 + L/\lambda_{eff})/N(d)$.

The contact resistance for the CNT with 12 nm average diameter has been measured at 8,859 Ω. The contact length between the CNT and the metal is typically approximately 1 µm in this configuration. We estimate that the specific contact resistivity between the CNT and the Ti/Au metal electrode is of the order of 10 µΩ cm^2. This value is close to the contact resistivity between the CNT and the Ti/Au electrode reported by others [AVO 05].

3.3.2. *CNT microwave transistors*

Scaling factor requirements in existing devices increased performances and density are faced to technological, economic and fundamental physics limits. For active components, an evolutionary approach based on well-established three-terminal transistor concepts using different materials, such as CNTs, has been suggested for the last 10 years to address many of the problems present in strong scaled semiconductor devices, for example limitations in transition frequency limit with gate length reduction (Figure 3.6).

This strategy intervenes with technology solutions of the near future, compared to solutions of the more distant future, suggesting new concepts such as two-terminal molecular devices, spintronics and quantum computing.

Compared to alternative materials tested by the silicon electronics industry, such as SOI, SiGe and high-k dielectrics, SWCNTs have some unique advantages including their nanoscale dimensions as well as electronic and optical properties stated from their band structure.

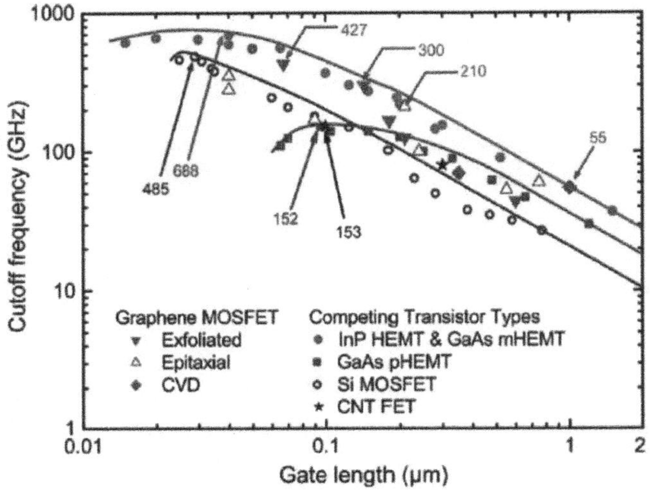

Figure 3.6. *Transition frequency predictions as a function of gate length for classical semiconductor and CNT-based transistor devices*

3.3.2.1. *State of the art of CNT-FETs components*

Because of their excellent transport properties, in terms of electron mobility and saturated velocity, and also of thermal dissipation (Table 3.2), CNTs have been first implemented in Si-p-type-based FETs structures to ensure free holes carrier transport between source and drain electrodes.

$$E_{Gap} = \frac{|t|.a_{C-C}}{d_{CNT}}$$

In addition, the CNT energy gap Egis controlled by CNT diameter d CNT according to [3.4] where a C-C is the distance between two Carbon atoms equal to 0.142 nm

Several CNT-FET configurations with back-gate electrode made by a p^{++} Si substrate and separated from source and drain by a SiO_2 layer, with a CNT draping over Au/Ti metal electrodes or a CNT deposited under (Ti/Co) metal electrodes, have been investigated [AVO 05, SAI 98] in order to reduce contact resistance from 1,000 to 30 kΩ, inducing enhancement of on-state currents up to 1 μA and transconductance from 0.001 to 0.3 μS.

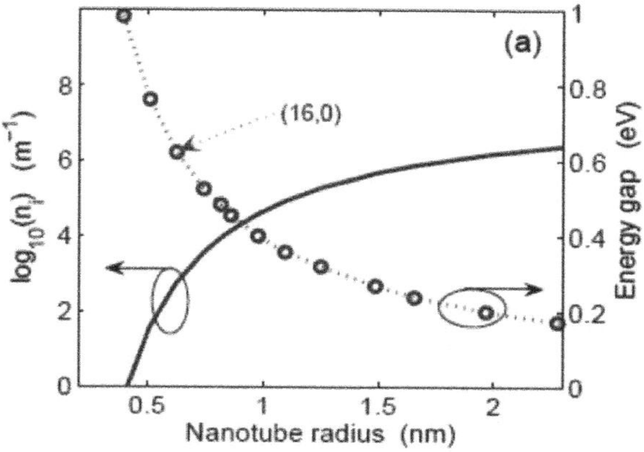

Figure 3.7. *CNT energy gap and intrinsic doping n_i as a function of tube radius*

	Bandgap (eV)	Electron Mobility (cm^2/Vs)	Saturated Electron Velocity (107 cm/s)	Thermal Conductivity (W/cm-K)
CNT	~0.9	100,000	>10	>30
InAs	0.36	33,000	0.04	0.27
Si	1.1	1,500	0.3	1.5
GaAs	1.42	8,500	0.4	0.5
InP	1.35	5,400	0.5	0.7
4HSiC	3.26	700	2.0	4.5
GaN	3.49	900	3.3	20

Table 3.2. *Comparison of CNT properties with other semiconductor materials [133]*

Top-gate CNT-FET components have been suggested [CAM 01] to reduce the threshold gate voltage from –12 to 0.5 V assumed by the gate dielectric layer thickness, improving also the gate-CNT channel coupling and the cutoff frequency (device capacitance reduction through the gate-to-source and gate-to-drain geometrical overlaps), and to improve transconductance to 3.25 μS. Optimizations of devices have been focused on the selection of suitable high dielectric gate insulator and electrode/CNT contact resistance metal [TAN 98, MAR 98, WIN 02].

Today, state-of-the-art devices from IEMN incorporate SWCNT samples containing 99% pure semiconducting SWCNTs (Figure 3.8), and achieve operating frequencies above 80 GHz [NOU 09].

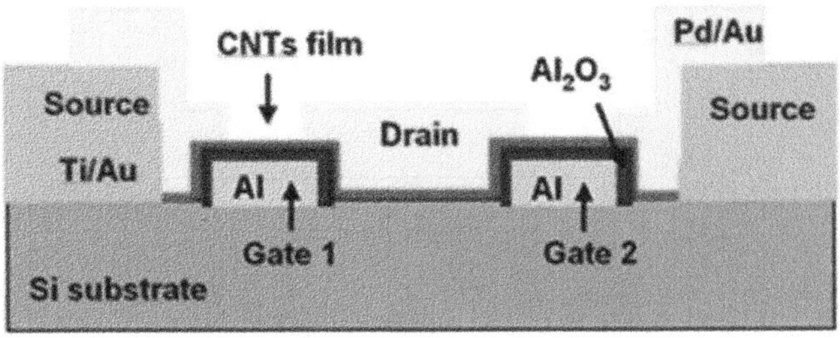

Figure 3.8. *Cross-sectional representation of state-of-the-art CNT FET on solid substrate (Si-based) [NOU 09]*

Recently, researchers at IBM proposed a new topology of transistors, owing to the deposit aligned films of CNT in solution, in a back-gate configuration. This approach is based on the dielectrophoresis method, which tries to avoid the migration of metal along the CNTs. They demonstrated extrinsic current gain and power gain cutoff frequencies, respectively, of 7 GHz and 15 GHz. While the extrinsic current gain is comparable to the state of the art, the extrinsic power gain is improved. The de-embedded, intrinsic current gain and power gain cutoff frequencies of 153 GHz and 30 GHz are the highest values experimentally achieved to date [STE 12].

3.3.3. *RF absorbing and shielding materials based on CNT composites*

Composites based on CNTs are very promising for the development of new shielding and absorbing materials at microwave frequencies with a high performance and a large operating frequency band, due to their high aspect ratio and special conductive properties, for an excellent absorption level with low CNT concentration [SAI 06].

CNTs have the potential to be used as nanoscale transmission lines and as high-performance passive components in nanoelectronics. In this chapter, we have discussed SWCNTs that were developed using a chemical vapor deposition technique. Both the DC and AC two-terminal resistances were measured and analyzed. The transmission properties of the SWCNTs were shown by comparing the S parameters of a device with CNTs with those of a device without a CNT. The RF characterization of the CNTs was carried out using the network analyzer at 20 GHz frequency [ZHA 06].

3.3.4. *Interconnects*

Metallic CNT bundles have been studied as potential candidates for next-generation interconnect conductors due to their excellent electrical properties and high electromigration resistance. Aligned CNT bundles offer advantages over randomly developed nanotubes in terms of lower resistance and easier process integration. Vertically aligned CNT bundles have been demonstrated as interconnect contact/via. The low resistance of 0.6 Ω for 2 μm CNT via has been achieved. Horizontally aligned CNT bundles are promising candidates for interconnect lines. In addition, we measured the global resistance of the CNT bundles. The intershell interaction within a single CNT is the second-order effect in our measurement compared to the global resistance. Therefore, it is reasonable to assume that the inner walls of the CNT will not contribute to the overall conductance. Assuming one CNT has one conducting graphite shell, the overall resistance of a single CNT is still much larger than the resistance of a 1D quantum wire R_0 (13 kΩ). This additional resistance is believed to result from the imperfect contact between the CNT and the metal or the phonon scattering in the diffusive region [CHA 10].

4

Nanotechnology-based Subsystems

Chapter 4 is devoted to nanotechnology-based subsystems. It will address the following topics: functions of sampling and analog-to-digital conversion, optical mixing of microwave signals, the field of nanoantennas including the optical control of antennas, THz photoconducting antennas, 2D and 1D material-based antennas. Challenges for future applications and reconfigurable antennas will also be analyzed.

4.1. Sampling and analog-to-digital converter

Analog-to-digital converters (ADCs) are required in numerous applications. As described in Figure 4.1, the requirements in terms of resolution (numbers of bits) are broadly distributed over a large range of frequencies depending on the applications, from consumer electronics to military and satellite systems.

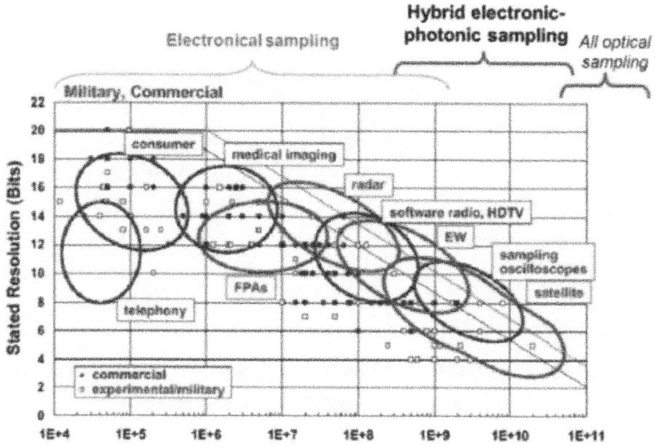

Figure 4.1. *Commercial and military applications of high-speed ADCs [WAL 99]. For a color version of this figure, see www.iste.co.uk/tripon/nanotechnology.zip*

The main limitations of high-speed ADCs are described in Figure 4.2. It shows the state of the art of purely electronic ADCs exhibiting the constraints in time aperture and theoretical limits.

Some specific parameters of the sampling or subsampling schemes are defined as follows:

1) *SINAD*: signal-to-noise and distortion ratio

– This represents the ratio of the power in the desired signal to the power in all the other spectral components.

2) *SNR*: signal-to-noise ratio

– This represents the ratio of the power in the desired signal to the power in all the other components, excluding the first five harmonics and DC.

3) *SFDR*: spurious-free dynamic range

– This defines the ratio of the power in the desired signal to the power in the next highest discrete spectral component.

4) Sampling jitter

– This is the variation in sampling time referred to the average clock rate.

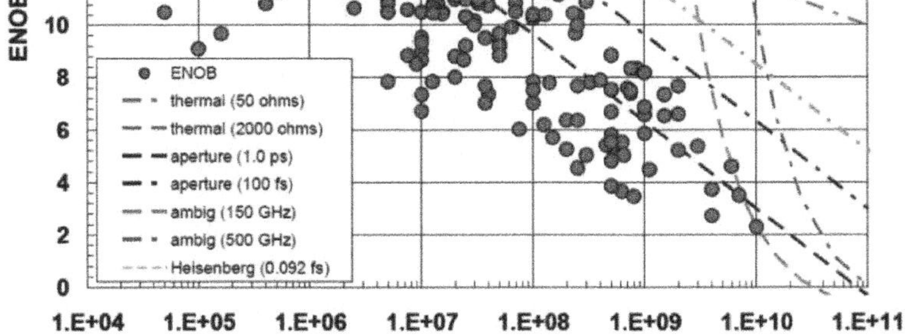

Figure 4.2. *High-speed ADC limitations and future prospects [WAL 08]. For a color version of this figure, see www.iste.co.uk/tripon/nanotechnology.zip*

4.1.1. *Basic principles of sampling and subsampling*

An example of an ideal sampling in the time and frequency domains is shown in Figure 4.3.

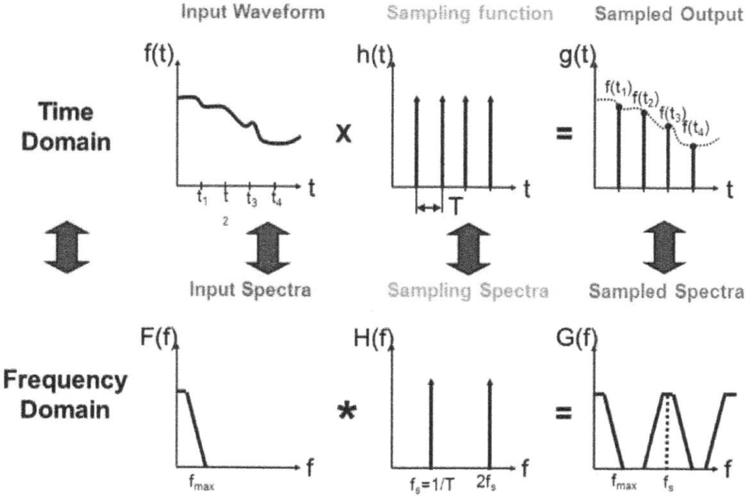

Figure 4.3. *Ideal sampling scheme*

A signal $f(t)$, after having been digitized at the frequency f_s, has a spectral representation equal to the spectral representation of $f(t)$ shifted every f_s. If $f_s > 2f_{\text{max}}$, there is no spectrum overlapping, where f_s is equal to $2 \cdot f_{\text{max}}$ defining the Nyquist frequency.

An example of an ideal subsampling in the time and frequency domains is shown in Figure 4.4. In this configuration, the input signal must have a defined bandwidth. It can be reconstructed only if $f_s > 2\text{BW}$.

One of the main requirements for digitizing high-frequency signals is associated with the jitter requirements defined in Figure 4.5, as has been summarized in the Chazelas "rule of ten" (the sampling of a 10 GHz signal with 10 bits of effective number of bits (ENOB) requires a jitter t in the range of 10 fs).

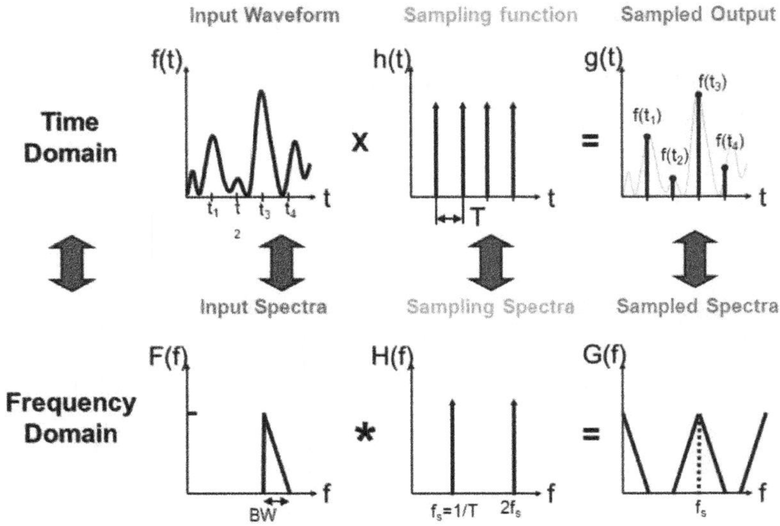

Figure 4.4. *Ideal subsampling scheme*

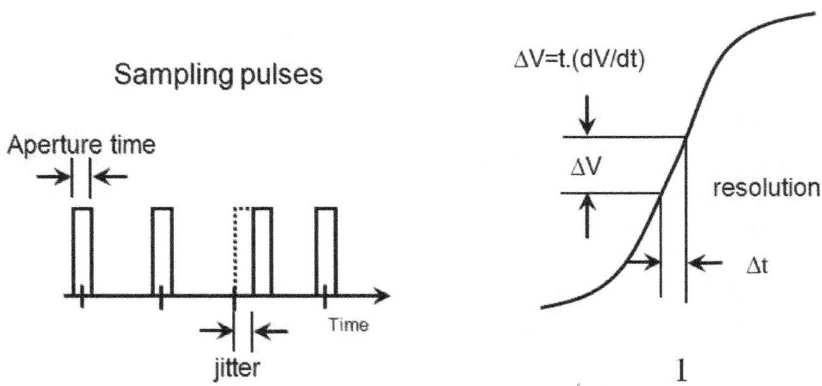

$$t = \frac{1}{2\pi f_s . 2^n}$$

Figure 4.5. *Time jitter definition*

4.1.2. *Optical sampling of microwave signals*

The approach described concerns the optical sampling of microwave signals [ROU 11, TRI 11, TRI 13]. The approach has focused on the following aspects:

1) The constraint of the jitter of the optical source controlling the sample devices, which defines the performances of the sampling system.

2) The simplicity of the switch structure included in the sampling function.

3) The linearity of the sampling system.

For applications requiring us to simultaneously handle wide bandwidth signals and high linearity, according to the Chazelas rule deduced from [4.1] for a sinusoidal incoming signal:

$$t = \frac{1}{2.\pi.f_{in}.2^n} \qquad [4.1]$$

where t represents the clock jitter, n is the desired effective number of bits and f_{in} is the maximum frequency of the incoming signal.

The constraints on the performances of the clock in terms of jitter, time aperture and repetition rate are extremely high. It shows that the jitter must be in 10 s of the fs region.

The proposed approach is based on a microwave switch that is optically controlled by a pulsed laser source (picosecond pulse width) (Figure 4.7) and high repetition rate. The simplest version of the photoswitch is a microstrip or coplanar interrupted transmission line (Figure 4.6).

4.2. Photomixing principle

Photomixers are heterodyne optoelectronic devices designed for the generation of ultra-high-frequency electromagnetic waves. These frequencies could reach the THz field based on the mixing of two light waves of slightly different wavelengths. Usually, these wavelengths are in the visible or IR region. The conversion principle is based on the detection of the temporal interference generated by the two optical beams [DID 62], and the frequency difference of the two lightwaves corresponds to the targeted microwave up to THz frequency.

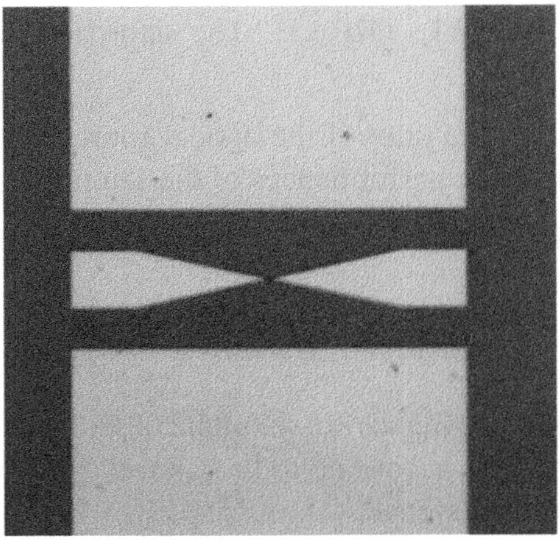

Figure 4.6. *Example of an ultra-high-speed wideband microwave photoconductive switch*

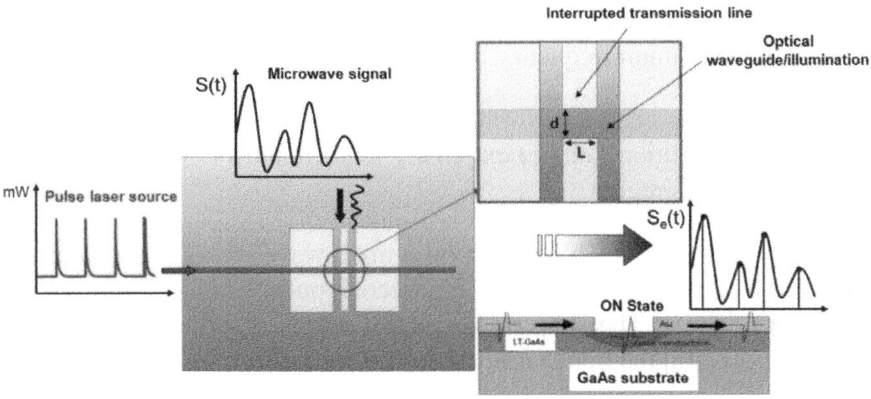

Figure 4.7. *Principle of the sampling architecture*

The active part of the photomixer is made of a semiconductor element with a resistivity modulated at ultra-high frequency. This modulation comes from the quadratic detection of the temporal interference between the two optical sources. The device is based on ultrashort lifetime photoconductive materials (~1 ps) compatible with the THz signal generation. Metal–semiconductor–metal structures are usually used [CHO 92].

The core of the submillimeter wave or THz photomixer is a photoconductive antenna, i.e. a dipole based on an ultra-high-speed material able to synchronize the carriers with the beating frequency (in the picosecond range based on LT-GaAs material, for example) [AUS 84].

4.3. Nanoantennas for microwave to THz applications

4.3.1. *Optical control of antennas in the microwave domain*

The photoconductive effect exploited in an RF device induces a phase shifting of the RF signal flowing in the device as shown in the imaginary part of the ON/OFF ratio in equation [1.19]. In this configuration, the envisaged phase variation depending directly on the incident optical power allows the use of the photoswitch as an optically controlled phase shifter.

This new functionality could in the future be inserted into phased array antennas requiring a phase and amplitude control of the transmitted microwave signals in order to control the radiation pattern of the antenna in the two dimensions.

4.3.2. *THz photoconducting antennas*

In the following, an array of photoconducting antennas that generate electrically controllable millimeter wave and submillimeter wave radiations in free space is described. Under quasi-sinusoidal optical illumination, the emitted radiation is directional and electrically steerable, and it can be scanned through an angle of more than 40°. The center of the scanning range can be adjusted by changing the angle of incidence of the pump beam. The far-field radiation pattern of the array is measured and discussed. Also, some properties of the array are demonstrated under illumination by a short optical pulse. These include electrical control of the frequency content of the signal and mapping of the spatial profile of the applied electric field onto the radiated waveform. The latter property can be used to multiplex information presented in parallel into a train of radiated pulses with a bit spacing of 6 ps.

4.3.3. *2D material-based THz antennas*

The electromagnetic properties of graphene have led to numerous research activities from an electromagnetic point of view; graphene can be described as an infinitesimally thin medium characterized by a surface conductivity [LOV 12, LLA 12, GOM 12].

A particularly promising emerging field is graphene-enabled wireless communications. Wireless communications among nanosystems cannot be achieved by simply reducing the size of a classical metallic antenna down to a few micrometers, since this would impose the use of very high resonant frequencies in the optical range. Due to the expected very limited power of nanosystems, the low mobility of electrons in metals when nanoscale structures are considered, and the challenges in implementing optical nanotransceivers [RUS 10, CHE 13], the feasibility of wireless communications among nanosystems would be compromised if this approach were followed. However, due to its groundbreaking properties, graphene technology has been widely studied to implement wireless communications among nanosystems.

Indeed, graphene-based plasmonic nanoantennas, a few micrometers in size, have been predicted to radiate electromagnetic waves at the THz band [LLA 12]. The important role of graphene conductivity in the characteristics of graphene antennas has been analyzed, and their radiation performance both in transmission and reception is studied numerically.

4.3.4. *1D material-based antennas*

Despite several international laboratories having theoretically studied the possibility of using nanotubes as antennas [HAN 05, BUR 06, FIT 07, HUA 08] in different frequency domains, such as THz [WAN 08, WU 08, CHO 10, JOR 10], GHz [KOK 10] and optics [CIU 10], no experimental demonstration has been performed to the best of our knowledge. Before going through the physics of nanoantennas, we briefly introduce the basics of antennas and then of nanotubes/nanowires separately. Thus, we will go from the macro-wire antenna model to the nanowire antenna model and review the similar and different parameters between both scales. Wire antennas are the oldest and most versatile antennas suited for various applications. It is a simple device and easy to understand most of the radiation mechanisms and the dipole structure simplification of radiating elements. The typical configuration is made up of two conductor wires, with a length of λ/2, as shown in Figure 4.8(a). The current distribution in the conductor wires (Figure 4.8(b)) can be considered in one dimension, due to the geometry (usually the *z*-axis direction). This is the time variation of the current distribution that will generate a radiated electromagnetic field in the surrounding space. Maxwell's equations lead to the relationship between the current variation $I(z)$ and the radiated field E_θ in the far-field space [4.2] [BAL 05]:

$$E_\theta = i\eta \frac{ke^{-ikr}}{4\pi r} \sin\theta \left[\int_{-l/2}^{+l/2} I(z) e^{ikz\cos\theta} dz \right] \quad [4.2]$$

where η is the characteristic impedance of free space, k is the constant propagation, l is the dipole length and r and θ are the radius and elevation angle coordinates, respectively. The 3D emission pattern is said to be omnidirectional because it only depends on θ. Figure 4.8(b) shows the typical radiation pattern of a wire antenna in a plane containing the *z*-axis.

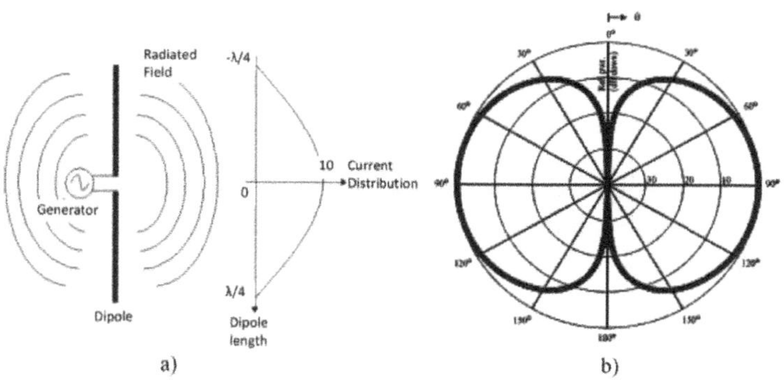

Figure 4.8. *a) Current distribution and free-space radiated field of a half-wavelength nanodipole and b) associated E-plane typical wire antenna pattern*

Along with the radiation pattern is a set of other key parameters that are used to quantify an antenna and its performances [BAL 05]:

1) The input impedance is the impedance that the power input circuit will have to match in order to transmit the maximum power to the radiating device.

2) The gain is the ratio of intensity, in a given direction, to the radiation intensity that would be obtained if the power accepted by the antenna was

radiated isotropically. The radiation intensity corresponding to the isotropically radiated power is equal to the input power accepted by the antenna divided by 4π.

3) The directivity is the ratio of the radiation intensity in a given direction from the antenna to the radiation intensity average over all directions.

4) The radiation efficiency is the ratio of the radiated power over the accepted power and is a dimensionless combined factor of both the conduction efficiency (losses through metal conduction) and dielectric efficiency (losses through propagation in dielectric space).

Size reduction toward nanometric scale changes the electromagnetic properties of the conducting elements. When a wire is fabricated whose cross-sectional dimension is comparable with the quantum mechanical (Fermi) wavelength of the electron, the wire essentially forms a single-mode waveguide for the electron waves. Then, in a 1D conductor such as a nanotube, the electrons are only free to move along the length of the wire, and not in the transverse direction. Therefore, the current distribution is effectively 1D. In addition to the electron transport occurring in only one dimension, we also have two more important effects: large resistance and large inductance.

While copper is typically used in applications where high conductivity is required, it does not maintain its bulk conductivity when scaled to nanometer dimensions. In contrast, nanotubes have better conductivity than copper when scaled to their diameter. It has recently been shown that the DC resistance per unit length of a single-walled CNT at room temperature is approximately 6 kΩ/µm. A copper wire with the same diameter (1.5 nm) would have an even higher resistance per unit length. This resistance per unit length is quite large compared with the characteristic impedance of free space, as well as typical radiation resistances in traditional antennas. Therefore, it cannot be neglected. Recently, it has been proven that the AC and DC resistances are the same for a nanotube up to approximately 10 GHz. However, these high impedances could be significantly reduced to 50 Ω if, instead of resistive contact between the CNT and the dielectric substrate, a capacitive contact is used [BUR 04]. This means that between the metallic contact and CNT, a thin dielectric is introduced. This particular configuration of the contact, i.e. metal/dielectric/nanotube, means that in parallel with the 6.5 kΩ resistance will be a capacitor with the overall effect of reducing the impedance of the CNT. Thus, the nano-T/R module could work at an impedance that is approximately 50 Ω, conferring compatibility with existing wireless systems.

The conductivities of nanotube and metallic wires are different because in a metallic wire the charges are relatively free of movement. This flux of charge is concentrated on the surface of the conductor in what is called the "conductivity skin depth". Due to the unique structure of carbon nanotubes, there is very little possibility for electrons to move in the same manner as in a macro-metal wire. In the case of nanotubes, the electron movement is made by ballistic transport through the nanotubes with a path length of approximately 100 nm in the tubular structure or via tunneling across gaps with an associated high tunneling resistance.

The first and most apparent change in characteristics is the wave propagation velocity and the resonance velocity. In a macro-model, the resonance wave velocity is equal to the speed of light, but in a nanotube antenna, it goes otherwise. As stated earlier, the wave propagation velocity in a nanotube transmission line is already 0.02 c. When used as a resonant dipole, the wave resonance in the nanotube can be associated with plasmons by the transmission line developed in [PIE 10b], where the propagation velocity of the antenna was found to be vp = 3vF = 0.01 c [PIE 10a]. However, this is only a theoretical rough approximation of the reality. Further calculation and experimentation give values of approximately 0.015 c and 0.017 c. The difference in wave propagation velocity for a metallic wire and a nanotube NT antenna is illustrated in Figure 4.9, by the difference in current distribution along the antenna.

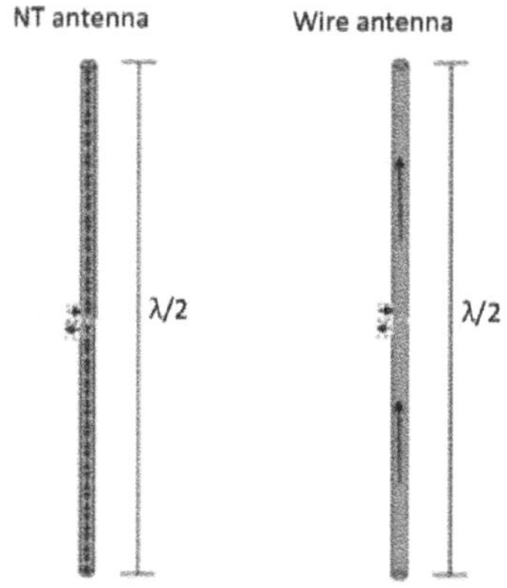

Figure 4.9. *Current distribution for nanotube versus wire antennas of the same length*

Some recent papers deal with the theoretical study of the interaction of 1D electronic systems with microwave radiation, leading to a quantitative theory of CNT antenna performance. Since CNTs can be developed having lengths in the order of centimeters, and can be metallic, a natural topic is to consider CNTs for centimeter and millimeter wave antenna applications. In fact, one of the technical issues that scientists have to face to bring nanotechnologies to reality is the communication and data exchange between the nano-sized devices or organisms and the macro-world. Using nanoantennas based on CNTs for wireless communications could solve this issue.

4.3.5. *Challenges for future applications*

4.3.5.1. *THz wireless communications*

Due to the increasing demand for wireless communication, requirements for more bandwidth and consequently an implied increase in carrier frequency for communications and data transfers have been deployed.

The main interests in the field of THz wireless communications have been extensively described, recalling some basic parameters such as the ultralarge bandwidth capacity compared with microwave systems, the better directionality and the "security" associated with the ultralarge bandwidth [FED 10]

Compared to IR systems, THz radiations have lower attenuation under constraint atmospheric conditions such as fog. THz radiation appears as a viable solution for the last mile and first mile problems [SAB 09a, COH 10] (the last and first mile problems refer to establishing broadband, multiuser local wireless connections with high-speed networks, i.e. fiber-optical).

Basic considerations of THz and sub-THz communication are discussed including:

1) free space versus guided waves;

2) directionality;

3) scintillations;

4) atmospheric and free-space damping including fog, rain and dust/smoke;

5) indoor versus outdoor communication.

4.3.5.1.1. Secure wireless THz communications

Many papers in the THz scientific literature claim that THz can be used for "secure" communications. Scenarios for secure links might include stealthy short-distance communications between vehicles – manned or unmanned – and personnel. Unmanned vehicles may require short-distance secure communication links so that they can receive instructions/transmit data before dispersing to conduct their remote-controlled or autonomous mission. For the link to be secure, unauthorized personnel should not be able to identify either the data via eavesdropping on communication channels or the presence of a communication link.

The main characteristics of THz communication for secure communications are:

1) highly directional beams compared to microwave communications;

2) less scattering of radiation compared to IR wireless;

3) limited propagation distance due to atmospheric attenuation;

4) encryption of the beam;

5) large channel bandwidth for spread spectrum techniques that enable anti-jamming and low probability of detection systems.

4.3.5.2. *Reconfigurable antennas*

This field appears as a very hot topic in the research of reconfigurable antennas and especially in the field of reconfigurable metamaterials for antenna applications.

Vardaxoglou from Loughborough University, UK studied in 2006 the impact of the use of photoconductive switches on the reconfiguration of microwave antennas opening the possibility of switching to all the required frequencies, thus eliminating the need for complicated wideband and multiband antenna solutions.

He studied an optically reconfigurable CPW-fed coplanar stripline (CPS) dipole antenna with an optically induced frequency shift of 39.4%, which is achieved from switches in the ON state to switches in the OFF state. The antenna shows ideal forward-plane radiation patterns and good boresight gain during switching operations. The performances of the phase shifters were evaluated using illumination from laser light-emitting diodes (LEDs) running at various optical power levels, varied by changing the driving current to the LEDs. In his experiments, the insertion loss with the LEDs ON can almost be maintained near the resonance frequency, and can be as low as 0.7 dB at 4.2 GHz.

Recently, Tripon-Canseliet and Maci studied a new concept of reconfigurable metasurface based on photoconductive switching.

Conclusions and Perspectives

In conclusion, we will draw some perspectives of this new field of optically controlled low dimensionality materials. It will focus on Van der Waals hetero-structures and "beyond graphene" materials as an example of nanoarchitectronics.

C.1. Conclusions

As discussed in this book, converged elements from recent research results show that microwave systems, technology and material-based architectures at the nanoscale lead to a novel approach and a novel scientific area.

As defined in the Introduction to this book, we propose to call this new area *nanoarchitectronics* due to its ability to mix material research and electromagnetic research, leading to the requirements to develop multiscale approaches.

Our analysis was focused on potential breakthrough technologies for future EM applications ranging from RF to THz due to the enhancement of quantum effects in 2D and 1D structures [GEI 13].

As an example and following the research on 2D atomic crystals, it appears extremely powerful to assemble isolated atomic planes into designer heterostructures made layer by layer in a precisely chosen sequence. In order to add new functionalities to graphene, research has been carried out on the modification of the graphene band structure, or its combination with other semiconductors in hybrid devices. A promising route toward making such devices is the use of graphene in conjunction with atomic layers of transition metal dichalcogenides (MX2) and gallium chalcogenides (GaX), or by creating graphene heterostructures with thin films of III-V semiconductors.

As an example, in the framework of the European Flagship Graphene, it is proposed to address the investigation of atomically thin 2D systems, including the growth and study of atomically thin 2D crystals beyond graphene; the production of graphene heterostructures with semiconductor materials; the investigation of their electronic and optical properties; the development of applications of hybrid systems in functional devices, and the growth and preparation of new layered systems for spintronics.

The application of this approach will cover the following topics:

– real-world applications in functional electronics or high-end instrumentation development (e.g. detectors and sensors);

– new device concepts that will change the currently available technologies and production methods in industrial sectors such as electronics, energy and sensors;

– stretching the fundamental limits of miniaturization in widely used devices.

C.2. Perspectives: beyond graphene structures for advanced microwave functions

Carbon-based nanostructures, such as carbon nanotubes and graphene sheets, have interesting electrical properties, which make it possible for carbon to play the same role as silicon and III-IV alloys have in the electronics revolution over the last five decades.

On the contrary, non-carbon-based nanostructures, such as gold, silver and zinc oxide nanoparticles, have important properties in the area of the optical and infrared range. These properties are known as plasmonics properties due to the interaction of electron gas in the crystals of these materials with electromagnetic fields in this spectrum range.

These nanostructures can be used individually as in the case of field effect transistors, nanoantennas or nanotransmission lines. They can also be used as composite structures as in the case of absorbing materials, shielding screens or composite substrates. In composite structures, nanomaterials play a significant role in modifying the properties of the host material.

C.2.1. van der Waals heterostructures

These heterostructures, called van der Waals heterostructures (Figure C.1), have recently been investigated, and their unusual properties and new phenomena have been revealed. Research into graphene and other 2D atomic crystals is intense and is likely to remain one of the leading topics in condensed matter physics and materials science for many years. Looking beyond this field, isolated atomic planes can also be reassembled into designer heterostructures made layer by layer in a precisely chosen sequence [GEI 13].

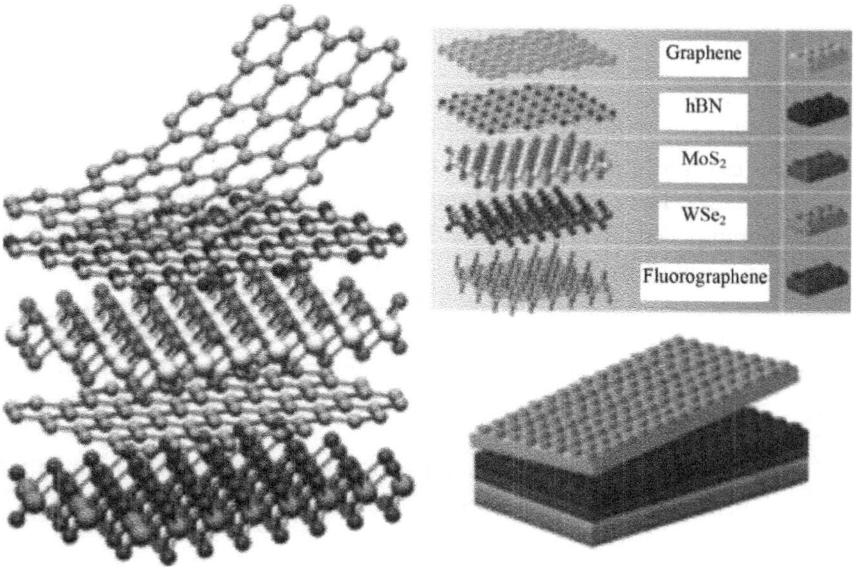

Figure C.1. *Example of van der Waals heterostructures from [GEI 13]. For a color version of this figure, see www.iste.co.uk/tripon/nanotechnolgoy.zip*

Based on Van Hove singularities as explained in Chapter 1, a new range of 2D heterostructures have been studied to add photoactive properties by using semiconducting transition metal dichalcogenides (TMDC)/graphene stacks and enhanced light–matter interactions [BRI 13], leading to enhanced photon absorption and electron-hole creation (which are collected in transparent graphene electrodes). This makes it possible to develop extremely efficient flexible photovoltaic devices with photoresponsivity above 0.1 A/W.

Graphene and hexagonal boron nitride (h-BN) have the same crystal structure and have very similar lattice constants, but unlike graphene, h-BN is an insulator with a large energy bandgap of 6 eV. Most previous studies have focused on the use of thick layers of BN as a substrate for graphene electronics or as a dielectric in experiments on coupled 2D electron gases.

Hexagonal boron nitride [BRI 12] is an appealing substrate because it has an atomically smooth surface that is relatively free of dangling bonds and charge traps. It also has a lattice constant similar to that of graphite, and has large optical phonon modes and a large electrical bandgap. High-quality exfoliated mono- and bilayer graphene devices on single-crystal h-BN substrates, using a mechanical transfer process, have been reported [DEA 10].

The nanosheet and nanowire heterostructure provide an architecture to integrate the extraordinary properties of 1D nanowires and 2D nanosheets into a 3D space. They hold great potential in applications that request complex nanomaterials with multiple functionality, high surface area and efficient charge transport.

For instance, this heterostructure can be useful for solar energy as photovoltaics and energy storage, such as supercapacitors and lithium ion batteries.

The epitaxial NS–NW heterostructure can also be very useful for other important fields, such as 3D optoelectronics and sensing [CHU 13].

The most widely studied 2D material is graphene. Engineering of a graphene bandgap increases the fabrication complexity and either reduces the mobilities to the level of strained silicon films or requires high voltages. Although single layers of molybdenum di-sulfate (MoS_2) have a large intrinsic bandgap of 1.8 eV, the recently reported mobilities were too low for practical devices. Approaches such as the use of a hafnium oxide gate dielectric were demonstrated at room temperature with of at least 200 cm^2/V s, similar to that of graphene nanoribbons. Monolayer MoS_2 could also be substituted for graphene in applications that require thin transparent semiconductors, such as optoelectronics and energy harvesting single-layer MoS_2 transistors.

C.2.2. Beyond graphene: heterogeneous integration of graphene with other 2D semiconductor materials

Silicon atoms crystallize in bonding configurations that are similar to those in graphene, opening up the potential for exceptional electronic properties with possible applications in nanoscale electronics.

Silicene [RAI 13], the graphene equivalent for silicon, could open new trends, providing new perspectives for application, especially due to its compatibility with Si-based electronics.

Since the discovery of graphene and the tremendous growth in this field of research, much effort has been made to search theoretically and experimentally for similar 2D materials composed of group IV elements, especially silicon [PAT 12]. However, silicene does not seem to exist in nature nor is there any solid phase of silicon similar to graphite. As a result, pure 2D silicene layers cannot be generated by exfoliation methods as performed initially in the case of graphene. More sophisticated methods have to be considered for the growth or synthesis of silicene. One promising concept for synthesizing silicene is to deposit silicon on metal surfaces that do not interact strongly with the Si atoms or that do not form compounds.

In order to develop real 2D silicene sheets, Ag(111) surfaces with six-fold top-layer symmetry [LEB 09] can be utilized, which might support the formation of a honeycomb Si layer.

Similar to silicene, research activities have also been devoted to germanene [BIA 13].

C.2.3. *Graphene allotropes*

Many 2D periodic carbon allotropes, e.g. graphyne and graphdiyne, can be envisioned [MAL 12, HIR 10]. Graphynes and graphdiynes are built from triple- and double-bonded units of two carbon atoms. Examples are given in Figure I.1. Many graphynes and graphdiynes including those depicted in Figure C.2 can be assumed to be chemically stable.

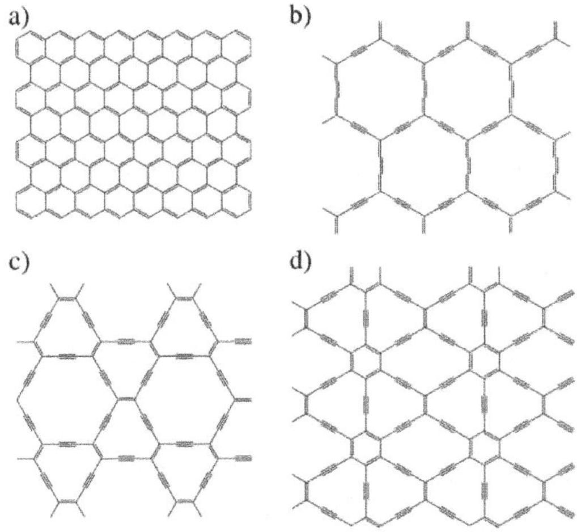

Figure C.2. *Structures of graphene and graphynes: a) graphene, b) α–graphyne, c) β–graphyne and d) 6,6,12-graphyne. In all cases, there is only one resonance structure, i.e. one of several equivalent Lewis structures*

All these areas of "beyond graphene" enhance the need for nanoarchitectronics simulation and design tools, methods, growth and technologies, which will be the main area in physics from our point of view over the next 10 years.

Bibliography

[ALA 09] ALAGARASI A., "Introduction to nanomaterials," in VISWANATHAN B. (ed.), *Nanomaterials*, Narosa Publishing House, 2009.

[ALU 07] ALU A., SILVEIRINHA M.G., SALANDRINO A., *et al.*, "Epsilon-near-zero metamaterials and electromagnetic sources: tailoring the radiation phase pattern", *Physical Review B*, vol. 75, 1555410, 2007.

[AMI 95] AMIRTHARAJ P.M., SEILER D.G., "Optical properties of semiconductors", *Handbook of Optics*, McGraw Hill, 1995.

[AND 89] ANDERSSON I., ENG S.T., "Phase and amplitude characteristics of InP:Fe modified interdigitated gap photoconductive microwave switches", *IEEE Transactions on Microwave Theory and Techniques*, vol. 37, no. 4, pp. 729–733, April 1989.

[AND 13] ANDRYIEUSKI A., LAVRINENKO A.V., "Graphene metamaterials based tunable terahertz absorber: effective surface conductivity approach", *Optics Express*, vol. 21, no. 7, pp. 9144–9155, 2013.

[ATT 12] ATTIYA A.M., "From nanotechnology in RF and microwave applications: review article", *29th National Radio Science Conference (NRSC 2012)*, Cairo University, Egypt, 2012.

[AUS 83] AUSTON D.H., "Impulse response of photoconductors in transmission lines", *IEEE Journal of Quantum Electronics*, vol. QE-19, no. 4, pp. 639–648, April 1983.

[AUS 84] AUSTON D.H., CHEUNG K.P., SMITH P.R., "Picosecond photoconductive Hertzian dipoles", *Applied Physics Letters*, vol. 45, pp. 284–286, 1984.

[AVO 05] AVOURIS P., RADOSAVLJEVIC M., WIND S.J., "CNT electronics and optoelectronics", in ROTKIN S.V., SUBRAMONEY S. (eds), *Applied Physics of Carbon Nanotubes: Fundamentals of Theory, Optics and Transport Devices, NanoScience and Technology*, Springer, Berlin/Heidelberg, pp. 227–251, 2005.

[AVO 10] AVOURIS P., "Graphene: electronic and photonic properties and devices", *Nano Letters*, vol. 10, pp. 4285–4294, 2010.

[BAL 05] BALANIS C.A., *Antenna Theory: Analysis and Design*, 3rd ed., Wiley-Interscience, 2005.

[BIA 13] BIANCO E., BUTLER S., JIANG S., *et al.*, "Stability and exfoliation of germanane: a germanium graphane analogue", *ACS Nano*, vol. 7, pp. 4414–4421, 2013.

[BRI 12] BRITNELL L., GORBACHEV R.V., JALIL R., *et al.*, "Atomically thin boron nitride: a tunnelling barrier for graphene devices", *Nano Lett.*, vol. 12, no. 3, pp. 1707–1710, 2012.

[BRI 13] BRITNELL L., RIBEIRO R.M., ECKMANN A., *et al.*, "Strong light matter interactions in heterostructures of atomically thin films", *Science 14*, vol. 340, no. 6, June 2013.

[BRY 06] BRYLLERT T., WERNERSSON L.E., FROBERG L.E., *et al.*, "Vertical high-mobility wrap-gated InAs nanowire transistor", *IEEE Electron Device Letters*, vol. 27, no. 5, pp. 323–325, May 2006.

[BUR 04a] BURKE P.J., "Lüttinger liquid theory as a model of the gigahertz electrical properties of Carbon nanotubes (erratum)", *IEEE Transactions on Nanotechnologies*, vol. 3, no. 2, pp. 331–331, March 2004.

[BUR 04b] BURKE P.J., "AC performances of nanoelectronics: towards a THz nanotube transistor", *Solid State Electronics*, vol. 40, no. 10–11, pp. 1981–1986, 2004.

[BUR 06] BURKE P.J., LI S., YU Z., "Quantitative theory of nanowire and nanotube antenna performance", *IEEE Transactions on Nanotechnology*, vol. 5, no. 4, pp. 314–334, July 2006.

[CAN 03] CANSELIET C., ALGANI C., DESHOURS F., *et al.*, "Contrôle optique d'antennes à balayage par photoconduction – Application à un réseau d'antennes intégrées sur GaAs", *Optique Hertzienne et Diélectrique,* Calais, France, September 2003.

[CAS 76] CASTAGNÉ R., LAVAL S., LAVAL R., "Picosecond 1-wavelength optoelectronic gate", *Electronics Letters*, vol. 12, no. 17, pp. 438–439, August 1976.

[CAS 06] CASTRO NETO A.H., GUINEA F., PERES N.M.R., "Drawing conclusions from graphene", *Physics World*, vol. 19, pp. 1–5, November 2006.

[CAS 09] CASTRO NETO A.H., GUINEA F., PERES N.M.R., *et al.*, "The electronic properties of graphene", *Reviews of Modern Physics*, vol. 81, pp. 109–162, 2009.

[CHA 08a] CHAN C.K., PENG H.L., LIU G., *et al.*, "High-performance lithium battery anodes using silicon nanowires", *Nature Nanotechnology*, vol. 3, no. 1, pp. 31–35, January 2008.

[CHA 08b] CHANG P.-C., LU J.G., "ZnO nanowire field-effect transistors", *IEEE Transactions on Electron Devices*, vol. 55, no. 11, pp. 2977–2987, November 2008.

[CHA 10] CHAI Y., XIAO Z., CHAN P.C.H., "Horizontally aligned carbon nanotube bundles for interconnect application: diameter-dependent contact resistance and mean free path", *Nanotechnology*, vol. 21, pp. 235705–235709, 2010.

[CHA 11] CHA S.N., CHOI J.H., BAIK C.W., *et al.*, "Perspectives on nanotechnology for RF and terahertz electronics", *IEEE Transactions on Microwave Theory and Techniques*, vol. 59, no. 10, pp. 2709 – 2718, 2011.

[CHE 10] CHEN H., XI N., LAI K.W.C., *et al.*, "Development of infrared detectors using single carbon-nanotube-based field-effect transistors", *IEEE Transactions on Nanotechnology*, vol. 9, no. 5, pp. 582–589, September 2010.

[CHE 13] CHEN P.-Y., ALU A., "All-graphene terahertz analog nanodevices and nanocircuits", *7th European Conference on Antennas and Propagation (EUCAP'13)*, Gothenburg, Sweden 2013.

[CHE 97] CHEN E., CHOU S.Y., "Characteristics of coplanar transmission lines on multilayer substrates: modeling and experiments", *IEEE Transactions On Microwave Theory and Techniq*ues, vol. 45, no. 6, June 1997.

[CHO 10] CHOI S., *et al.*, "Design of efficient terahertz antennas: CNT versus gold", Apsursi, pp. 1–1, Toronto, 2010

[CHO 92] CHOU S.Y., LIU M.Y., "Nanoscale terahertz metal-semiconductor-metal photodetectors", *IEEE Journal of Quantum Electronics*, vol. 28, no. 10, pp. 2358–2368, 1992.

[CIU 10] CIU X., *et al.*, "Sphere on pillar optical nano-antennas", *IEEE Nanotechnology Materials and Devices Conference (NMDC)*, vol. 171, pp. 171–176, 2010.

[COH 10] COHEN M.L., ZETTL A., "The physics of boron nitride nanotubes", *Physics Today*, pp. 34–38, November 2010.

[COM 05] COMINI E., FAGLIA G., SBERVEGLIERI G., *et al.*, "Tin oxide nanobelts electrical and sensing properties," *Sensors and Actuators B: Chemical*, vol. 111, pp. 2–6, November 2005.

[COM 08] COMBRIÉ S., DE ROSSI A., TRAN Q., *et al.*, "GaAs photonic crystal cavity with ultrahigh Q: microwatt nonlinearity at 1.55 µm", *Optical Letters*, vol. 33, pp. 1908–1910, 2008.

[COS 09] COSCIA U., AMBROSONE G., AMBROSIO A., *et al.*, "Photoconductivity of multiwalled CNT deposited by CVD", *Solid State Sciences*, vol. 11, no. 10, pp. 1806–1809, 2009.

[COU 08] COUTAZ J.L., (ed.), *Optoélectronique THz*, EDP Sciences, 2008.

[CUI 03] CUI Y., ZHONG Z.H., WANG D.L., *et al.*, "High performance silicon nanowire field effect transistors", *Nano Letters*, vol. 3, no. 2, pp. 149–152, February 2003.

[CUR 08] CURUTCHET A., THÉRON D., WERQUIN M., *et al.*, "Nonlinear characterization and modeling of carbon nanotube field-effect transistors", *IEEE Transactions on Microwave Theory and Techniques*, vol. 56, no. 7, pp. 1505–1510, July 2008.

[DAI 11] DAI X., WANG J., LARRUE A., *et al.*, "GaAs nanowire field-effect-transistors for RF applications", *International Conference on Materials for Advanced Technologies (ICMAT 2011)*, Singapore, 2011.

[DE 06] DE CAMARGO E CASTRO L., Modeling of carbon nanotube field-effect transistors, PhD Thesis, University of British Columbia, 2006.

[DE 81] DE SALLES A.A.A., "Optical control of GaAs MESFET's", *IEEE Transaction on Microwave Theory & Technology*, vol. MTT-31, pp. 812–820, 1981.

[DEA 10] DEAN C.R., YOUNG A.F., MERIC I., *et al.*, "Boron nitride substrates for high-quality graphene electronics", *Nature Nanotechnology*, vol. 5, pp. 722–726, 2010.

[DEC 11] DECROSSAS E., El-GHAZALY S.M., "Microwave dielectric characterization of carbon nanotube networks", in BIANCO S. (ed.), *Materials Science, Composite Materials "Carbon Nanotubes: From Research to Applications"*, Intech, 2011.

[DEC 12] DECROSSAS E., El SABBAGH M. A., HANNA V. F., et al., "Rigorous characterization of carbon nanotube complex permittivity over a broadband of RF Frequencies", *IEEE Transactions on Electromagnetic Compatibility*, vol. 54, pp. 81–87, February 2012.

[DID 62] DIDOMENICO M., SVELTO Jr. O., PANTELL R.H. *et al.*, "Optical frequency mixing in bulk semiconductors", *Applied Physics Letters*, vol. 1, no. 4, pp. 77–79, 1962.

[DRA 07] DRAGOMAN M., DRAGOMAN D., MULLER A.A., "High frequency devices based on graphene", *International Semiconductor Conference (CAS '07)*, pp. 53–56, Sinaia, Romania 2007.

[DRE 05] DRESSELHAUS M.S., DRESSELHAUS G., SAITO R., *et al.*, "Raman spectroscopy of carbon nanotubes", *Physics Reports*, vol. 409, pp. 47–99, 2005.

[ENG 06] ENGHETA N., ZIOLKOWSKI R.W. (eds), *Metamaterials: Physics and Engineering Explorations*, John Wiley & Sons, New York, 2006.

[ENG 07] ENGHETA N., "Circuits with light at nanoscales: optical nanocircuits inspired by metamaterials", *Science*, vol. 317, pp. 1698–1702, September 2007.

[EL 09a] EL SABBAGH M.A., EL-GHAZALY M.A.,"Miniaturized carbon nanotube-based RF resonator", *IEEE MTT-S International Microwave Symposium Digest*, Boston, MA, 7–12 June 2009.

[EL 09b] EL SABBAGH M.A., EL-GHAZALY S.M., NASEEM H.A., "Carbon nanotube-based planar transmission lines", *IEEE MTT-S International Microwave Symposium Digest*, Boston, MA, 7–12 June 2009.

[EL 09c] EL SABBAGH M.A., EL-GHAZALY S.M., "Frequency-dependent circuit models of carbon nanotube networks", *Electrical Performance of Electronic Packaging and Systems*, Portland, OR, 19–21 October 2009.

[EST 09] ESTACIO E., *et al.*, *34th International Conference on Infrared, Millimeter, and Terahertz Waves*, Busan, pp. 773–774, 21–25 September 2009.

[FAC 05] FACI S., *et al.*, "Modélisation électrique non linéaire de photocommutateurs microrubans pour la conception de circuits microondes actifs et passifs commandés par voie optique", *14ème Journées Nationales Microondes*, Nantes, France, May 2005.

[FED 10] FEDERICI J., MOELLER L., "Review of terahertz and subterahertz wireless communications", *Journal of Applied Physics*, vol. 107, pp. 111101, 2010.

[FEK 11] FEKECS A., BERNIER M., MORRIS D., *et al.*, "Fabrication of high resistivity cold-implanted InGaAsP photoconductors for efficient pulsed terahertz devices," *Optical Materials Express*, vol. 1, pp. 1165–1177, 2011.

[FIT 07] FITCHTNER N., *et al.*, "Investigation of copper and carbon nanotubes antennas using thin wire integral equations", *IEEE Asia Pacific Microwave Conference (APMC)*, Asia-Pacific, 2007.

[FRE 03] FREITAG M., MARTIN Y., MISEWICH J.A., *et al.*, "Photoconductivity of single carbon nanotubes", *Nano Letters*, vol. 3, no. 8, pp. 1067–1071, 2003.

[FRI 13] FRIEDLEIN R., FLEURENCE A., OZAKI T., *et al.*, "Silicene: atom-thick silicon with tunable properties", *SPIE Newsroom*, 7 June 2013.

[FRO 92] FROBERG N.M., HU B.B., ZHANG X.-C., AUSTON D.H., *et al.*, "Terahertz radiation from a photoconducting antenna array", *Quantum Electronics, IEEE Journal of*, vol. 28, no. 10, pp. 2291–2301, October 1992.

[GAN 04] GANGLOFF L., MINOUX E., TEO K.B.K., *et al.* "Self-aligned, gated arrays of individual nanotube and nanowire emitters", *Nano Letters*, vol. 4, no. 9, pp. 1575–1579, July 29 2004.

[GAN 13] GAN X., SHIUE R.-J., GAO Y., *et al.*, "Chip-integrated ultrafast graphene photodetector with high responsivity", *Nature Photonics*, 15 September 2013.

[GEI 13] GEIM A.K., GRIGORIEVA I.V., "Van der Waals heterostructures", *Nature*, vol. 499, pp. 419–425, 2013.

[GOM 12] GOMEZ-DIAZ J.S., PERRUISSEAU-CARRIER J., "Microwave to THz properties of graphene and potential antenna applications", *Proceedings of ISAP*, 2012.

[GRA 12] GRAHAM C., GWILLIAM R., SEEDS A., "Nitrogen ion implanted InP based photo-switch", *Optics Express*, vol. 20, no. 2, November 2012.

[GUO 06] GUO Y., *et al.*, "Theoretical investigation on photoconductivity of single intrinsic carbon nanotubes", *Applied Physics Letters*, vol. 88, pp. 133111, 2006.

[HAN 05] HANSON G.W., "Fundamental transmitting properties of carbon nanotube antennas", *IEEE Transaction on Antennas and Propagation*, vol. 53, no. 11, pp. 3426–3435, November 2005.

[HAQ 06] HAQUE S., MARINELLI C., UDREA F., *et al.*, "Absorption characteristics of single wall carbon nanotubes", *NSTI Nanotechnology Conference*, vol. 1, pp. 134–137, 2006.

[HAT 11] HATEM O., CUNNINGHAM J., LINFIELD E.H., *et al.*, "Terahertz-frequency photoconductive detectors fabricated from metal-organic chemical vapor deposition-grown Fe-doped InGaAs", *Applied Physics Letters*, vol. 98, no. 12, pp. 121107, 2011.

[HER 02] HERTEL T., *et al.*, "Charge-carrier dynamics in single-wall carbon nanotube bundles: a time-domain study", *Applied Physics A*, vol. 75, pp. 449–465, 2002.

[HES 10] HESHMAT B., *et al.*, "Evaluation of carbon nanotubes for THz photomixing", *Radar Conference*, Washington DC, pp. 1176–1179, May 2010.

[HOC 08] HOCHBAUM A.I., CHEN R.K., DELGADO R.D., *et al.*, "Enhanced thermoelectric performance of rough silicon nanowires", *Nature*, vol. 451, no. 7175, pp. 163–167, January 2008.

[HUA 08] HUANG Y., *et al.*, "Performance predication of carbon nanotubes bundle dipole antenna", *IEEE Transactions on Nanotechnology*, vol. 7, pp. 331–337, 2008.

[IIJ 91] IIJIMA S., "Helical microtubules of graphitic carbon", *Nature*, vol. 354, pp. 56–58, 1991.

[JAV 13] JAVEY A., "2D semiconductors: materials, interfaces, and devices", *224th ECS Meeting*, The Electrochemical Society, San Francisco, CA, 2013.

[JOH 05] JOHANSSON A., WIDENKVIST E., LU J., "Fabrication of high-aspect-ratio prussian blue nanotubes using a porous alumina template", *Nano Letters*, vol. 5, pp. 1603–1606, 2005.

[JOR 10] JORNET J.M., *et al.*, "Graphene-based nano-antennas for electromagnetic nanocommunications in the terahertz band", *Proceedings of the 4th European Conference on Antennas and Propagation (EuCAP)*, Barcelona, Spain, 12–16 April 2010.

[KAS10] KASSEM H., VIGNERAS V. LUNET G., "Characterization techniques for materials' properties measurement", in MININ I. (ed.), *Microwave and Millimeter Wave Technologies from Photonic Bandgap Devices to Antenna and Applications*, 2010.

[KIN 10] KINOSHITA M., STEINER M., ENGEL M., *et al.*, "The polarized carbon nanotube thin film LED", *Optics Express*, vol. 18, pp. 25738–25745, 2010.

[KLE 13] KLEKACHEV A.V., NOURBAKHSH A., ASSELBERGHS I., *et al.*, "Graphene transistors and photodetectors", *The Electrochemical Society Interface*, Spring 2013.

[KOK 10] KOKSAL C.E., *et al.*, "Design and analysis of systems based on RF receivers with multiple CNT antennas", *Nano Communication Networks*, vol. 1, pp. 160–172, 2010.

[LEB 09] LEBÈGUE S., ERIKSSON O., *Physical Review B*, vol. 79, pp. 115409–1, 115409–4, 2009.

[LEY 09] LEYMAN R., CARNEGIE D., BAZIEVA N., *et al.*, "Characterisation of InAs:GaAs quantum dot-based photoconductive THz Antennas", *34th International Conference on Infrared, Millimeter, and Terahertz Waves*, pp. 773–774, September 2009.

[LEY 13] LEYMAN R., CARNEGIE D., BAZIEVA N., *et al.*, "Characterisation of InAs:GaAs quantum dot-based photoconductive THz antennas", *IEEE Photonics Conference (IPC)*, pp. 418–419, 2013.

[LI 13] LI C., YU Y., CHI M., *et al.*, "Epitaxial nanosheet-nanowire heterostructures", *Nano Letters*, vol. 13, pp. 948–953, 2013.

[LID 92] LIDEIKIS T., NAUDZIUS K., TREIDERIS G., *et al.*, "Picosecond GaAs and InGaAs photocnductive switches obtained by low-temperature meta-organic chemical vapour deposition", *Semiconductor Science and Technology*, vol. 7, pp. 845–849, 1992.

[LIN 02] LIN W.Z., LIU Z.G., LLAO R., *et al.*, "Characterization of femtosecond low-temperature grown GaAs photoconductive switch", *Chinese Physics Letters*, vol. 19, no. 4, pp. 557–559, 2002.

[LIN 97] LIN G.-R., PAN C.-L., "Picosecond responses of low dosage arsenic-ion-implanted GaAs photoconductors", *Applied Physics Letters*, vol. 71, no. 20, pp. 2901–2903, November 1997.

[LIU 05] LIU F., BAO M., WANG K.L., *et al.*, "One dimensional transport of In2O3 nanowires," *Applied Physics Letters*, vol. 86, no. 21, pp. 213–101, May 2005.

[LIU 09] LIU Y., *et al.*, "Alignment enhanced photoconductivity in single wall carbon nanotube films", *Nanotechnlogy*, vol. 20, no. 8, pp. 035203, 2009.

[LLA 12] LLATSER I., KREMERS C., CHIGRIN D.N., *et al.*, "Radiation characteristics of tunable graphennas in the terahertz band", *Radioengineering*, vol. 21, no. 4, pp. 946–953, 2012.

[LOV 12] LOVAT G., "Equivalent circuit for electromagnetic interaction and transmission through graphene sheets", *IEEE Transactions on Electromagnetic Compatibility*, vol. 54, no. 1, 2012.

[MAK 08] MAK K.F., SFEIR M.Y., WU Y., *et al.*, "Measurement of the optical conductivity of graphene", *Physical Review Letters*, vol. 101, pp. 196405–1, 196405–4, 2008.

[MAN 07] MANGENEY J., CHIMOT N., MEIGNIEN L., *et al.*, "Emission characteristics of ion-irradiated In0.53Ga0.47As based photoconductive antennas excited at 1.55 µm", *Optics Express*, vol. 15, pp. 8943–8950, 2007.

[MAR 98] MARTEL R., SCHMIDT T., SHEA H.R., *et al.*, *Applied Physics Letters*, vol. 73, pp. 2447–2449, 1998.

[MAR 03] MARTENSSON T., BORGSTROM M., SEIFERT W., *et al.*, *Nanotechnology*, vol. 14, pp. 1255–1258, 2003.

[MCI 97] MCINTOSH K.A., NICHOLS K.B., VERGHESE S., *et al.*, "Investigation of ultrashort photocarrier relaxation times in low-temperature-grown GaAs", *Applied Physics Letters*, vol. 70, no. 3, pp. 354–356, 1997.

[MÉL 04] MÉLIN T., DIESINGER H., DERESMES D., *et al.*, "Probing nanoscale dipole-dipole interactions by electric force microscopy", *Physics Review Letters*, vol. 92, pp. 166101, 2004.

[MER 08] MERIC I., BAKLITSKAYA N., KIM P., *et al.*, "RF performance of top-gated, zero-bandgap graphene field-effect transistors," *IEDM Technical Digest*, pp. 1–4, 2008.

[MEY 05] MEYYAPAN M., *Carbon Nanotubes: Science and Applications*, 1st ed., CRC Press LLC, FL, 2005.

[MIN 01] MINAMI N., KAZAOUI S., JACQUEMIN R., *et al.*, "Optical properties of semiconducting and metallic single wall carbon nanotubes: effects of doping and high pressure", *Synthetic Metals*, vol. 116, pp. 405–409, 2001.

[MIN 04] MINOUX E., TEO K.B.K., VINCENT P., *et al.*, "Self-aligned, gated arrays of individual nanotube and nanowire emitters", *Nano Letters*, vol 4, n°9, pp 1575–1579, July 29, 2004

[NG 04] NG H.T., HAN J., YAMADA T., *et al.*, "Single crystal nanowire vertical surround-gate field-effect transistor", *Nano Letters*, vol. 4, pp. 1247, 2004.

[NIL 08] NILSSON H.A., DUTY T., ABAY S., *et al.*, "A radio frequency single-electron transistor based on an InAs", *Nano Letters*, vol. 8, no. 3, pp. 872–875, March 2008.

[NOU 09] NOUGARET L., HAPPY H., DAMBRINE G., *et al.*, "80 GHz field-effect transistors produced using high purity semiconducting single-walled carbon nanotubes", *Applied Physics Letters*, vol. 94, pp. 243505–1, 243505–3, 2009.

[NOV 04] NOVOSELOV K.S., GEIM A.K., MOROZOV S.V., *et al.*, "Electric field effect in atomically thin carbon films", *Science*, vol. 306, pp. 666–669, 2004.

[OBE 11] OBENG Y., SRINIVASAN P., "Graphene: is it the future for semiconductors?", *The Electrochemical Interface*, pp. 47–52, Spring 2011.

[ORT 01] ORTIZ V., STELLMACHER M., MARCADET X., *et al.*, "Control of low-temperature-grown GaAs for ultrafast switching applications", vol. 4280, *Proceedings of the SPIE*, 23 pp. 202–210, April 2001.

[PAC 09] PACEBUTAS V., BERTULIS K., ALEKSEJENKO G., *et al.*, "Molecular-beam-epitaxy grown GaBiAs for terahertz optoelectronic applications", *Journal of Materials Science: Materials in Electronics*, vol. 20, pp. S363–S366, 2009.

[PAS 08] PASSACANTANDO M., BUSSOLOTTI F., GROSSI V., *et al.*, "Photoconductivity in defective carbon nanotube sheets under ultraviolet–visible near infrared radiation", *Applied Physics Letters*, vol. 93, no. 5, pp. 051911–051913, 2008.

[PED 04] PEDERSEN T.G., "Exciton effects in carbon nanotubes", *Carbon*, vol. 42, nos. 5–6, pp. 1007–1010, 2004.

[PEN 99] PENDRY J.B., HOLDEN A.J., ROBBINS D.J., *et al.*, "Magnetism from conductors and enhanced nonlinear phenomena", *IEEE Transactions on Microwave Theory and Techniques*, vol. 47, no. 11, pp. 2075–2084, November 1999.

[PEN 07] PENDRY J.B., "Metamaterials and the control of electromagnetic fields", *Conference on Coherence and Quantum Optics OSA Technical Digest (CD) (Optical Society of America)*, paper CMB2, 2007.

[PIE 10a] PIERANTONI L., "RF nanotechnology–concept, birth, mission, and perspectives", *IEEE Microwave Magazine*, vol. 11, no. 4, pp. 130–137, June 2010.

[PIE 10b] PIERANTONI L., COCCETTI F., "Radio-frequency nanoelectronics: a new paradigm in electronic systems design", *Proceedings of Asia-Pacific Microwave Conference*, Yokohama, Japan, 2010.

[POS 13] POSPISCHIL A., HUMER M., FURCHI M.M., *et al.*, "CMOS-compatible graphene photodetector covering all optical communication bands", *Nature Photonics*, pp. 892–896, 2013.

[QIN 08] QIN Y., WANG X.D., WANG Z.L., "Microfibre-nanowire hybrid structure for energy scavenging", *Nature*, vol. 451, no. 7180, pp. 809–813, February 14, 2008.

[QIU 05] QIU X., FREITAG M., PEREBEINOS V., *et al.*, "Photoconductivity spectra of single-carbon nanotubes: implications on the nature of their excited states", *Nano Letters*, vol. 5, pp. 749–752, 2005.

[RAD 11] RADISAVLJEVIC B., RADENOVIC A., BRIVIO J., *et al.*, "Single-layer MoS transistors", *Nature Nanotechnology*, vol. 6, pp. 147–150, 2011.

[RAF 04] RAFAILOV E.U., *et al.*, "Photonics technology letters", *IEEE*, vol. 16, pp. 2439–2441, 2004.

[RUB 94] RUBIO A., CORKHILL J.L., COHEN M.L., "Theory of graphitic boron nitride nanotubes", *Physical Review B*, vol. 49, no. 7, pp. 5081–5084, February 1994.

[ROS 55] ROSE A., "Performance of photoconductors", *Proceedings of the IRE*, vol. 43, no. 12, pp. 1850–1869, December 1955.

[ROS 98] ROSENCHER E., VINTER B., *Optoélectronique*, Masson, Paris, France, 1998.

[ROU 11] ROUX J.F., DELORD J.M., COUTAZ J.L., "RF frequency response of photoconductive samplers", *IEEE Journal of Quantum Electronics*, vol. 47, no. 2, pp. 223–229, 2011.

[RUS 10] RUSSER P., FICHTNER N., LUGLI P., *et al.*, "Nanoelectronics-based integrated antennas", *IEEE Microwave Magazine*, pp. 58–71, December 2010.

[RUT 09] RUTHERGLEN C., JAIN D., BURKE P., "Nanotube electronics for radiofrequency applications", *Nature Nanotechnology*, vol. 4, pp. 811–819, 2009.

[SAI 98] SAITO R., DRESSELHAUS G., DRESSELHAUS M.S., *Physical Properties of Carbon Nanotubes*, 1st ed., Imperial College Press, London, 1998.

[SAI 06] SAIB A., BEDNARZ L., DAUSSIN R., *et al.*, "Carbon nanotube composites for broadband microwave absorbing materials", *IEEE Transactions on Microwave Theory and Techniques*, vol. 54, no. 6, pp. 2745–2754, 2006.

[SAL 07] SALEH B.E.A., TEICH M.C., *Fundamentals of Photonics*, Wiley Interscience, 2007.

[SAR 08] SARTORIUS B., ROEHLE H., KÜNZEL H., *et al.*, "All-fiber terahertz time-domain spectrometer operating at 1.55 μm telecom wavelengths", *Optics Express*, vol. 16, no. 13, 9565–9570, June 2008.

[SCH 06] SCHROEDER D.K., *Semiconductor Material and Device Characterization*, Wiley, 2006.

[SCH 10] SCHWIERZ F., "Graphene transistors", *Nature Nanotechnology*, vol. 5, pp. 487–496, July 2010.

[SMI 73] SMITH W., "Effect of light on selenium during the passage of an electric current", *Nature*, p. 303, February 1873.

[SMI 00] SMITH D.R., PADILLA W.J., VIER D.C., *et al.*, "Composite medium with simultaneously negative permeability and permittivity", *Physical Review Letters*, vol. 84, no. 18, pp. 4184–4187, 2000.

[SOU 12] SOUNAS D.L., CALOZ C., "Gyrotropy and nonreciprocity of graphene for microwave applications", *IEEE Transactions on Microwave Theory and Techniques*, vol. 60, no. 4, pp. 901–914, 2012.

[STE 12] STEINER M., ENGEL M., LIN Y.-M., *et al.*, "High-frequency performance of scaled carbon nanotube array field-effect transistors", *Applied Physics Letters*, vol. 101, pp. 053123–4, 2012.

[STO 09] STOKES P., LIU L.W., ZOU J.H., *et al.*, "Photoresponse in large area multiwalled carbon nanotube, polymer nanocomposite films", *Applied Physics Letters*, vol. 94, pp. 042110–042113, 2009.

[SUN 08] SUN J.L., XU J., ZHU J.L., *et al.*, "Disordered multiwalled carbon nanotube mat for light spot position detecting", *Applied Physics A: Material Science & Processing*, vol. 91, no. 2, pp. 229–233, 2008.

[SUN 11] SUN Y., RUSLI, SINGH N., "Room-temperature operation of silicon single-electron transistor fabricated using optical lithography", *IEEE Transactions on Nanotechnology*, vol. 10, no. 1, pp. 96–98, 2011.

[TAN 96] TAN H.K., JAGADISH C., KORONA K.P., *et al.*, "Ion-implanted GaAs for subpicosecond optoelectronic applications", *IEEE Journal of Selected Topics in Quantam Electronics*, vol. 2, pp. 636–642, 1996.

[TAN 98] TANS S.J., VERSCHUEREN A.R.M., DEKKER C., "Room-temperature transistor based on a single carbon nanotube", *Nature*, vol. 393, no. 49, pp. 49–52, 1998.

[TAN 13] TAN K.H., YOON S.F., WICAKSONO S., *et al.*, "Low temperature grown GaNAsSb: a promising material for photoconductive switch application", *Applied Physics Letters*, vol. 103, no. 11, pp. 111113, 1–4, 2013.

[TRI 06] TRIPON-CANSELIET C., FACI S., BLARY K., *et al.*, "Optically-controlled microwave phase shifting and sampling by efficient photoconductive switching on LT-GaAs substrate integrated technology", *International Conference on Application of Photonic Technology*, Quebec, Canada, June 2006.

[TRI 11] TRIPON-CANSELIET C., FACI S., HAPPY H., *et al.*, "Optically controlled nanotechnologies for microwave integrated devices", *IEEE International NanoElectronics Conference (INEC)*, Taipei, Taiwan, 2011.

[TRI 12] TRIPON-CANSELIET C., FACI S., PAGIES A., *et al.*, "Microwave on/off ratio enhancement of GaAs photoconductive switches at nanometer scale, *IEEE Journal of Lightwave Technology*, vol. 1, no. 12, pp. 3576–3579, 2012.

[TRI 13] TRIPON-CANSELIET C., FACI S., DECOSTER D., *et al.*, "Nano photoconductive switches for microwave applications", *Proceedings of SPIE Photonics West*, San Francisco, CA, 2013.

[TRI 14] TRIPON-CANSELIET C., XAVIER S., MODREANU M., *et al.*, *NEMO 2014*, 2014.

[VAN 07] VAN ZEGHBROECK B., *Principles of Semiconductor Devices*, Colarado University, 2007.

[VEY 80] VEYRES C., HANNA V.F., "Extension of the application of conformal mapping techniques to coplanar lines with finite dimensions", *International Journal of Electronics*, vol. 48, pp. 47–56, 1980.

[VIT 12] VITIELLO M.S., VITI L., ROMEO L., *et al.*, "Semiconductor nanowires for highly sensitive, room-temperature detection of terahertz quantum cascade laser emission", *Applied Physics Letters*, vol. 100, p. 241101, 2012.

[VOG 12] VOGT P., DE PADOVA P., QUARESIMA C., *et al.*, "Silicene: compelling experimental evidence for graphene like two-dimensional silicon", *Physical Review Letters*, vol. 108, pp. 155501–1, 155501–5, 2012.

[WAL 08] WALDEN R.H., "Analog-to-digital converters and associated IC technologies", *IEEE CSIC*, pp. 1–2, 2008.

[WAL 99] WALDEN R.H., "Analog-to-digital converter survey and analysis", *IEEE Journal on Selected Areas in Communications*, vol. 17, no. 4, pp. 539–550, April 1999.

[WAN 08] WANG Y., *et al.*, "Properties of terahertz wave generated by the metallic carbon nanotube antenna", *Chinese Optics Letters*, vol. 6, no. 10, pp. 770–772, 2008.

[WAN 09a] WANG H., NEZICH D., KONG J., *et al.*, "Graphene frequency multipliers", *IEEE Electron Device Letters*, vol. 30, no. 5, pp. 547–549, May 2009.

[WAN 09b] WANG J., LEE C.H., BANDO Y., *et al.*, "Multiwalled boron nitride nanotubes: growth, properties, and applications", in YAP Y.K. (ed.), *B-C-N Nanotubes and Related Nanostructures*, Springer Science + Business Media, LLC, pp. 906–908, 2009.

[WAN 10a] WANG J., LEE C.H., YAP Y.K., "Recent advancements in boron nitride nanotubes", *Nanoscale*, vol. 2, pp. 2028–2034, 2010.

[WAN 10b] WANG H., HSU A., WU J., *et al.*, "Graphene-based ambipolar RF mixers", *IEEE Electron Device Letters*, vol. 31, no. 9, pp. 906–908, September 2010.

[WAN 12] WANG S., ZHANG Z.Y., PENG L.M., "Doping-free carbon nanotube optoelectronic devices", *Chinese Science Bulletin*, vol. 57, nos. 2–3, pp. 149–156, 2012.

[WAN 13] WANG X., CHENG Z., XU K., *et al.*, "High-responsivity graphene/silicon-heterostructure waveguide photodetectors", *Nature Photonics*, vol. 7, pp. 888–891, 2013.

[WEI 06] WEI J., SUN J., ZHU J., *et al.*, "Carbon nanotube macrobundles for light sensing", *Small*, vol. 2, nos. 8–9, pp. 988–993, 2006.

[WIK] WIKIPEDIA

[WIN 02] WIND S.J., APPENZELLER J., MARTEL R., *et al.*, *Applied Physics Letters*, vol. 80, pp. 3817–3819, 2002.

[WU 08] WU Q., *et al.*, "Terahertz generation in the carbon nanotubes antenna", *IEEE*, vol. 978, 2008.

[WU 12] WU Y., JENKINS K.A., VALDES-GARCIA A., *et al.*, "State-of-the-art graphene high-frequency electronics", *Nano Letters*, vol. 12, pp. 3062–3067, 2012.

[YAN 12] YANG H., HEO J., PARK S., *et al.*, "Graphene barristor, a triode device with a gate-controlled Schottky barrier", *Science*, vol. 336, pp. 1140–1143, 2012.

[YEN 77] YEN H.W., *et al.*, "Switching of GaAs IMPATT diode oscillator by optical illumination", *Applied Physics Letters*, vol. 31, no. 2, pp. 120–122, 1977.

[XIA 06] XIANG J., LU W., HU Y.J., *et al.*, "Ge/Si nanowire heterostructures as high-performance field-effect transistors", *Nature*, vol. 441, no. 7092, pp. 489–493, May 2006.

[ZHA 06a] ZHAO J., CHEN X., XIE J.R.H., "Optical properties and photonic devices of doped carbon nanotubes", *Analytica Chimica Acta*, vol. 568, nos. 1–2, pp. 161–170, May 2006.

[ZHA 06b] ZHANG M., HUO X., CHAN P.C.H., *et al.*, "Radio-frequency characterization for the single-walled carbon nanotubes", *Applied Physics Letters*, vol. 88, no. 16, pp. 163109–1631096-3, 2006.

[ZHA 10a] ZHAO Z.-Y., SCHWAGMANN A., OSPALD F., *et al.*, "1.55 µm photoconductive THz emitters based on ErAs:In0.53Ga0.47As superlattices", *Photonics Society Winter Topical Meeting Series (WTM)*, *IEEE*, 2010.

[ZHA 10b] ZHANG J., *et al.*, "Length-dependent photoinduced current in vertically-aligned MWNTs forests", *Microsystem Technologies*, vol. 16, pp. 2115–2118, 2010.

[ZHA 13] ZHANG R., ZHANG Y., ZHANG Q., *et al.*, "Growth of half-meter long carbon nanotubes based on Schulz–Flory distribution," *ACS Nano*, vol. 7, no. 7, pp. 6156–6161, June 2013.

[ZIA 08] ZIAEI A., BAILLIF M.L., DEMOUSTIER S., *et al.*, "Microwave applications of carbon nanotubes: nano-antennas and nanoswitches", in VARADAN V.K., (ed.), *Nanosensors and Microsensors for Bio-Systems*, *Proceedings of SPIE*, vol. 6931, 2008.

Index

M, N, O

P

S

T, Y, Z